ESO ASTROPHYSICS SYMPOSIA
European Southern Observatory

Series Editor: Jacqueline Bergeron

T0215127

Springer
Berlin
Heidelberg
New York
Barcelona
Hong Kong
London
Milan
Paris
Singapore
Tokyo

Physics and Astronomy **ONLINE LIBRARY**

http://www.springer.de/phys/

ESO ASTROPHYSICS SYMPOSIA
European Southern Observatory

Series Editor: Jacqueline Bergeron

Series homepage – http://www.springer.de/phys/books/eso/

A. Fitzsimmons D. Jewitt R.M. West (Eds.)

Minor Bodies
in the Outer Solar System

Proceedings of the ESO Workshop
Held at Garching, Germany,
2-5 November 1998

Springer

Volume Editors

Alan Fitzsimmons

Physics Department, Queen's University of Belfast
University Road
Belfast BT7 1NN, Northern Ireland/UK

David Jewitt

Institute for Astronomy, University of Hawaii
2680 Woodlawn Drive
Honolulu, HI 96822, USA

Richard M. West

European Southern Observatory
Karl-Schwarzschild-Strasse 2
85748 Garching, Germany

Series Editor

Jacqueline Bergeron

European Southern Observatory
Karl-Schwarzschild-Strasse 2
85748 Garching, Germany

Library of Congress Cataloging-in-Publication Data

Minor bodies in the outer solar system : proceedings of the ESO workshop held at
Garching, Germany, 2-5 November 1998 / A. Fitzsimmons. D. Jewitt, R.M. West (eds.).
   p. cm. -- (ESO astrophysics symposia)
   Includes bibliographical references.

   1. Kuiper Belt--Congresses. I. Fitzsimmons, A. (Alan) II. Jewitt, D. (David) III. West,
Richard M., 1941- IV. ESO Workshop on Minor Bodies of the Outer Solar System (1998
: Garching bei München, Germany) V. Series.

QB695 .M55 2001
523.2--dc21

                                                                00-049694

ISBN 978-3-642-07437-0   e-ISBN 978-3-540-40034-9

Springer-Verlag Berlin Heidelberg New York
a member of BertelsmannSpringer Science+Business Media GmbH

© Springer-Verlag Berlin Heidelberg 2010
Printed in Germany

Cover design: Erich Kirchner, Heidelberg

Preface

The recent discovery of the Kuiper Belt has revitalized the astronomical study of the Solar System and is beginning to open new and unexpected windows on the physics of planetesimal accretion. With more and better observational data being obtained at the technological limit of current facilities, a new perception of the relationships that exist among the various classes of small solar system body has emerged. Objects dislodged from the Kuiper Belt may be scattered into the region of the gas giant planets, where they are classified as Centaurs, and then passed to the region of the terrestrial planets to appear as Jupiter-family short-period comets. De-volatilized short-period comets may masquerade as near-Earth asteroids: a few so-called transition objects are already known.

This integrated view of the Solar System is highly revealing. We are beginning, for instance, to see possible links between structures in our own system and those present in dusty disks around nearby stars. The new observations have also motivated a number of fascinating theoretical studies in Solar System dynamics. Fundamental progress in the understanding of the Kuiper Belt rests heavily on access to moderate and large aperture telescopes. Physical studies, in particular, are still limited to the brightest objects even when using the largest (8-m to 10-m) diameter telescopes. For this reason, the Very Large Telescope (VLT) holds the potential to make a large impact on the future study of the minor bodies of the outer Solar System.

A four-day ESO Workshop on Minor Bodies of the Outer Solar System (ESO MBOSS-98) was held at the ESO Headquarters in Garching, Germany on November 2–5, 1998. The purpose of the meeting, attended by about 100 astronomers from around the globe, was to summarise the status of observational knowledge of the Kuiper Belt at that time, and to galvanize interest in using ESO's newly completed VLT for studies of Kuiper Belt Objects. Written versions of presentations from MBOSS-98 are included here. Light editing has been performed, mainly for purposes of clarification.

The organisers and participants are most grateful to Elisabeth Völk, Hans-Hermann Heyer, Ed Janssen and Hans-Jürgen Kraus for their efficient help with the organization and logistics of this workshop. Thanks are also due to Pamela Bristow for support with the preparation of the manuscript.

Belfast, Honolulu, Garching,
July 2000

Alan Fitzsimmons
Dave Jewitt
Richard West

Contents

Possible Mechanism of Cometary Outbursts 177
Subhon Ibadov

The Uppsala-DLR Trojan Survey
of the Preceding Lagrangian Cloud 179
Claes-Ingvar Lagerkvist, Stefano Mottola, Uri Carsenty,
Gerhard Hahn, Andreas Doppler, Arno Gnädig

The Distant Satellites of Uranus and the Other Giant Planets 187
Brian G. Marsden, Gareth V. Williams, Kaare Aksnes

List of Registered Participants

Name	Institution
ABE, Shinsuke	National Astronomical Observatory of Japan avell@pub.mtk.nao.ac.jp
AFONSO, Cristina	Centre d'Energie Atomique-CEA Cristina.Afonso@cea.fr
ARTYMOWICZ, Pawel	Stockholm Observatory pawl@astro.su.se
BAILEY, Mark	Armagh Observatory meb@star.arm.ac.uk
BARUCCI, Maria Antonietta	Observatoire de Paris – DESPA barucci@obspm.fr
BOEHNHARDT, Hermann	ESO, Chile hboehnha@eso.org
BROWN, Michael	University of Melbourne mbrowne@physics.unimelb.edu.au
COLAS, François	Bureau des Longitudes, Paris colas@bdl.fr
CONSOLMAGNO, Guy, SJ	Vatican Observatory gjc@specola.va
CREMONESE, Gabriele	Osservatorio Astronomico di Padova cremonese@pd.astro.it
DAVIES, John	Joint Astronomy Center – UKIRT jkd@jach.hawaii.edu
DELAHODDE, Catherine	ESO, Chile cdelahod@eso.org
DUNCAN, Martin	Queen's University, Ontario, Canada

FITZSIMMONS, Alan
Queen's University of Belfast
a.fitzsimmons@qub.ac.uk

FLETCHER, Edel
Queen's University of Belfast
E.Fletcher@qub.ac.uk

GABRYSZEWSKI, Ryszard
Space Research Centre
of Polish Academy of Sciences
kacper@cbk.waw.pl

GAJDOS, Stephan
Astronomical Institute, Bratislava
gajdos@fmph.uniba.sk

GLADMAN, Brett
Observatoire de Nice
gladman@obs-nice.fr

GOLDMAN, Bertrand
Centre d'Energie Atomique – CEA
Bertrand.Goldman@cea.fr

GREEN, Simon
University of Kent at Canterbury
s.f.green@ukc.ac.uk

GRUEN, Eberhard
Max-Planck-Institut fr Kernphysik
Eberhard.Gruen@mpi-hd.mpg.de

HAINAUT, Olivier
ESO, Chile
ohainaut@eso.org

IBADOV, Subhon
Institute of Astrophysics, Dushanbe
subhon@academy.td.silk.org

IPATOV, Sergei
Institute of Applied Mathematics,
Moscow
ipatov@spp.keldysh.ru

JEWITT, David
University of Hawaii
jewitt@ifa.hawaii.edu

KINOSHITA, Daisuke
Science University of Tokyo
daisuke@pub.mtk.nao.ac.jp

LAGERKVIST, Claes-Ingvar
Astronomiska Observatoriet
classe@astro.uu.se

LAZZARIN, Monica
Padova University
lazzarin@pd.astro.it

LUU, Jane
Leiden University
luu@cfa.harvard.edu

MARSDEN, Brian
Harvard-Smithsonian Center
for Astrophysics
bmarsden@cfa.harvard.edu

McBRIDE, Neil — University of Kent
N.McBride@ukc.ac.uk

MEECH, Karen — Institute for Astronomy, Hawaii
meech@ifa.hawaii.edu

MIYAMOTO, Atsushi — Saji Observatory
sajinet@infosakyu.ne.jp

NAKAMURA, Takashi — University of Electro Communications
nakamrta@cc.nao.ac.jp

PRIALNIK, Dina — Tel Aviv University
dina@planet.tau.ac.il

ROQUES, Françoise — Observatoire de Paris – DESPA
roques@obspm.fr

ROUSSELOT, Philippe — Observatoire de Besancon
philippe@obs-besancon.fr

SCHULZ, Rita — ESA/ESTEC
rschulz@estec.esa.nl

SEKIGUCHI, Tomohiko — National Astronomical Observatory of Japan
sekiguti@pub.mtk.nao.ac.jp

STERN, Alan — Southwestern Research Institute, Boulder
astern@swri.edu

TERRILE, Richard — Jet Propulsion Laboratory
rich.terrile@jpl.nasa.gov

THOMAS, Nicolas — Max-Planck-Institut fr Aeronomie
thomas@linax1.mpae.gwdg.de

TICHA, Jana — Klet Observatory
klet@klet.cz

TICHY, Milos — Klet Observatory
klet@klet.cz

TRUJILLO, Chad — University of Hawaii
chad@ifa.hawaii.edu

UNAL, Oktay — Ege University, Turkey
oktayun@astronomy.sci.ege.edu.tr

WATANABE, Jun-ichi — National Astronomical Observatory of Japan
watanabe@pub.mtk.nao.ac.jp

WEST, Richard M. — ESO, Garching
rwest@eso.org

The Kuiper Belt: Overview

David Jewitt

Institute for Astronomy, University of Hawaii, 2680 Woodlawn Drive, Honolulu, HI 96822, USA

Abstract. I briefly review the known and suspected properties of the Kuiper Belt, and pose several questions to which answers should be sought.

1 Introduction

The study of the Kuiper Belt has emerged as one of the leading subjects in planetary science. The Kuiper Belt offers many intriguing clues to the nature of the early solar system, and may also reveal key information about the processes behind the formation and growth of planets in the accretion disk of the sun. Kuiper Belt science has already been reviewed in detail several times elsewhere in the astronomical literature. In this short overview paper, I separate those aspects of the Kuiper Belt which are known with considerable confidence from other aspects that are less well known. In this way, I hope to provide a context for the other papers presented at the ESO Minor Bodies Workshop. To save space, and in the interests of keeping a tight focus on the subject at hand, I omit source references from this overview. The reader is directed to two full length reviews for complete references to the research literature (Jewitt, Annual Reviews of Earth and Planetary Sciences, 1999; Jewitt and Luu, Protostars and Planets IV, 2000). Both are available at http://www.ifa.hawaii.edu/faculty/jewitt/papers/.

2 Things We Know About the Kuiper Belt

2.1 Existence

The existence of the Kuiper Belt has been firmly established since the first direct observations of Kuiper Belt Objects (KBOs) starting in 1992. The possibility that objects might be present beyond Pluto was the subject of published speculation by Kenneth Edgeworth in 1943.

2.2 The name is unsatisfactory

Edgeworth's publications were not cited by his more famous contemporary Gerard Kuiper who, in 1951, wrote a review paper that included speculation about objects beyond Pluto (which he believed to be a massive planet in its own right). Perhaps in retribution, some have suggested that the Kuiper

Belt should instead be named after Edgeworth. The hybrid (but clumsy) Edgeworth-Kuiper Belt is occasionally used, as in the present Proceedings. In fact, neither Edgeworth nor Kuiper predicted the key features (numbers, sizes, masses, orbital properties) of the objects we now know to populate the outer solar system. Indeed, neither possessed a physical theory capable of making such predictions. For this reason the more descriptive appelation Trans-Neptunian Belt is sometimes used.

2.3 Sample statistics and population estimates

At the time of writing (March 1999) 115 KBOs have been reported. (This number was 320 in July 2000: Eds). Of these, approximately 50 have been observed at more than one opposition, and thus possess reasonably secure orbital elements. A complete and frequently updated list of KBOs is maintained by the Minor Planet Center at
http://cfa-www.harvard.edu/cfa/ps/lists/TNOs.html.

The cumulative sky-plane surface density, Σ $[deg^{-2}]$, can be fitted by a power law

$$log \Sigma = \alpha (m_R - m_o)$$

where α and m_o are constants and m_R is the apparent red magnitude. Published estimates of α vary from 0.58 to 0.7, while unit surface density is reached at $m_o = 23.3 \pm 0.1$. The cumulative surface density increases by a factor $10^{\alpha} \approx 4$ mag^{-1} to 5 mag^{-1}. The surface densities of KBOs brighter than $m_R = 21$ and fainter than $m_R = 26$ are not well constrained.

Most of the known objects have been discovered in narrow field-of-view surveys on telescopes of moderate (2-m) to large (10-m) aperture. Estimation of the Kuiper Belt population is subject to considerable uncertainty, due to bias effects in the data. Specifically, the surveys are flux limited, not volume limited, meaning that large and nearby KBOs are more readily detected than small and/or distant ones. The intrinsic population must thus be inferred from the survey data using models of the spatial and size distributions in the Kuiper Belt. From such models, the number of objects larger than diameter D = 100 km is found to be N(100km) $\sim 10^5$, in the 30 < R < 50 AU range. By extrapolation, N(5km) $\sim 10^9$, with an uncertainty of at least an order of magnitude.

2.4 Dynamical groups

The orbital elements of KBOs are divided into three distinct groups. The majority ($\approx 2/3$) are "Classical KBOs", with semi-major axes 42 < a < 50 AU and modest eccentricities ($e \sim 0.1$) that help them maintain a large separation from Neptune even when at perihelion. About 1/3 of the known objects are located in mean-motion resonances with Neptune (most in the 3:2 mean motion resonance at a = 39.4 AU). These are the resonant or "Plutino" objects, which seem to be dynamical counterparts of Pluto. A single object,

1996 TL_{66}, is neither Classical nor resonant, and is the prototype of the third dynamical category. The Scattered KBOs possess large semi-major axes, eccentricities and inclinations but have perihelia within $\simeq 5$ AU of Neptune, to which they are weakly dynamically coupled [20 are known as of July 2000 - Eds.].

2.5 Inclination distribution

The apparent distribution of orbital inclinations has FWHM = 10 ± 1 deg. Six KBOs have i > 30 deg, and one (1999 CY_{118}) has i > 40 deg. Observational selection effects discriminate against objects of high inclination and, therefore, the apparent FWHM must be a lower limit to the intrinsic width of the inclination distribution, which probably exceeds 30 deg.

2.6 Velocity dispersion

The root mean square inclinations and eccentricities of KBOs indicate a velocity dispersion $\Delta V = 1$ km/s. Collisions between all but the largest KBOs should be erosive rather than agglomerative.

2.7 Discovery distances

The distances at which KBOs are discovered fall in the range $26 < R < 48$ AU. The eight KBOs so far discovered with R < 30 AU are Neptune-crossing Plutinos.

2.8 Size distribution

The size distribution is consistent with a differential power law having index $q \approx -4$, with an uncertainty of about ± 0.5, in the diameter range $100 < D < 2000$ km. Smaller objects have not yet been sampled with observational confidence. The largest known KBO is Pluto (D = 2300 km). The largest discovered in the modern era is 1996 TO_{66} (D$\approx$ 800 km assuming surface albedo $p_R = 0.04$).

2.9 Mass

The mass in observable objects (D> 100 km) is of order $0.1 M_{Earth}$. The primary uncertainty in this estimate results from the unmeasured albedos of KBOs. The derived mass is proportional to $albedo^{-3/2}$. Crude dynamical limits to the Kuiper Belt mass of order $1 M_{Earth}$ have been placed based on the long-term stability of the orbit of comet P/Halley and on the absence of perturbations to spacecraft passing through the Belt.

2.10 Colors

The optical colors range from neutral to red, with a dispersion that exceeds the uncertainties of measurement. Compositional diversity, at least of the surface layers, is thus implied. The reddest KBOs are coated by a material that is not observed elsewhere in the solar system, except on the surfaces of some Centaurs (which are themselves recent escapees from the Kuiper Belt). Collisional resurfacing has been suggested as a possible cause of the color dispersion. According to this hypothesis, the instantaneous surface color results from a competition between reddening due to cosmic ray damage and resurfacing by fresh, unreddened material due to occasional impacts.

3 Things We Think We Know

3.1 Cometary reservoir

Dynamical considerations indicate that the short-period comets are more probably derived from a flattened source than from a spherical source like the Oort Cloud. It is suspected that the Kuiper Belt is this source. The inferred number of small KBOs, $N(5km) \approx 10^9$, is probably sufficient to supply the short-period comet flux over the age of the solar system. However, detailed comparisons between the cometary and Kuiper Belt populations are severely impaired by the near absence of direct measurements of the sizes of these objects. The median size of the nuclei of short-period comets, for example, is very poorly determined, while the sizes of KBOs are based on assumed low albedos. Neither is it clear from where in the Kuiper Belt the short-period comets originate, although chaotic zones near resonances and the scattered KBO population have both been suggested as sources.

3.2 Formation

It is likely that Kuiper Belt Objects formed before Neptune reached its current mass, since disturbances by Neptune would magnify the velocity dispersion among planetesimals in the adjacent disk and so impede collisional coagulation. The formation time for Neptune is uncertain, but possibly of order 10^8 yrs.

Models of planetesimal accumulation suggest that the present mass of KBOs is too small to allow formation of even 100 km sized objects in 108 yrs. Initial Kuiper Belt masses near $10 M_{Earth}$ to $30 M_{Earth}$ (i.e. 100 times the present mass) are needed to guarantee formation of KBOs up to and including Pluto-sized bodies in 10^8 yrs or less.

3.3 Dust

Impacts onto the Voyager 1 and 2 spacecraft when beyond 30 AU are plausibly ascribed to micron-sized dust particles generated in the Kuiper Belt.

The optical depth in micron-sized particles is $\tau \approx 10^{-7}$, about 4 orders of magnitude smaller than the estimated optical depth of the β Pictoris dust disk. The production rate of micron-sized dust particles implied by the Voyager detections is near 10^3 kg s^{-1}. This is close to the dust production rate expected from erosion of KBOs by interstellar dust particles.

3.4 Colors

There is evidence that the optical colors of KBOs might be bimodally distributed. Measurements between the optical and the near-infrared (J) band, however, show no evidence for bimodality. If confirmed, a bimodal distribution would eliminate collisional resurfacing as a plausible cause of color diversity in the Kuiper Belt.

3.5 Spatial distribution

The flux-limited longitudinal distribution of KBOs should vary with angular separation from Neptune. The Plutinos, for example, reach perihelion near ±90 deg from Neptune, and should be overabundant in these directions relative to others. Like Pluto, they may also occupy the "argument of perihelion libration", which maintains perihelion at high ecliptic latitude and therefore further minimizes Neptune perturbations. Longitudinal structure is apparent even in existing surveys but variations with latitude have not been adequately sampled.

3.6 Constraints from Pluto

The large specific angular momentum of the Pluto-Charon binary is often taken as evidence that Charon was created by a giant impact into Pluto. However, the specific angular momentum remains somewhat uncertain, pending accurate determinations of the density of Charon. The density of Pluto (about 2000 kg m^{-3}) indicates a large rock fraction, but it is unlikely that this density is representative of smaller KBOs.

3.7 Other Plutos

Models fitted to the measured sky-plane surface density of KBOs allow the possibility that other Pluto-sized objects await discovery in the Kuiper Belt. All published surveys, including Tombaugh's original, admit this possibility. Accretion models by Scott Kenyon and Jane Luu typically produce a handful of Pluto-sized objects, rather just a single example. More Plutos may soon be discovered.

4   Things We Would Like to Know

4.1   How were the resonances populated?

Radial migration of the planets (and their resonances) in response to torques exerted on the planetesimal disk can lead to trapping provided a) the migration is outward, smooth and slow and b) the planetesimal disk is initially cold (low inclination and eccentricity). As developed by Renu Malhotra, this attractive "resonance sweeping hypothesis" makes observationally verifiable predictions about the structure of the Kuiper Belt. The most eccentric Plutinos ($e \approx 0.34$) indicate that Neptune migrated radially by ≈ 7 AU, possibly on a timescale of a few million years.

If Neptune did not migrate appreciably, or if its drift were inwards rather than outwards, then the resonant populations are less easily understood. It has been suggested that KBOs, weakly scattered by massive interlopers, were captured in proportion to the small fraction of phase space occupied by resonances.

4.2   What was the initial mass of the Kuiper Belt?

As remarked above, the long growth times of KBOs suggest formation in a disk containing perhaps 100 times more mass than in the present Kuiper Belt. A similar factor is obtained by simple extrapolation of the smeared surface density of the planetary disk, as determined from the masses of the major planets. Are these estimates realistic, or is some crucial aspect of planetesimal agglomeration missing from our characterization of the problem?

4.3   How was the mass lost?

Collisional grinding has been suggested but not yet realistically modelled. For example, even at the present inclinations and eccentricities, the larger (diameter > 100 km) KBOs cannot be collisionally shattered. What was the distribution of collisional velocities in the epoch when collisions were supposedly dominant? The Plutinos, which are very weakly bound to the 3:2 mean-motion resonance, do not look like a population that has been heavily collisionally modified. Indeed, collisions would preferentially eject these objects from resonance.

Ejection by massive (Earth-mass?) projectiles has been suggested but, to date, not explored by means of a realistic numerical simulation. Could massive interlopers clear 99% of the mass of the primordial Kuiper Belt? What specific, observationally verifiable predictions are made by this hypothesis?

A 100 times more massive Kuiper Belt would presumably generate a flux of short-period comets 100 times that presently observed. Is there evidence in the early cratering records of the surfaces of planetary bodies for this elevated flux? Could the prolonged terminal bombardment phase of the inner solar system's first 0.5 Gyr be due in part to clearing of the massive Kuiper Belt?

4.4   How were the classical orbits excited?

The same hypothetical Earth-mass projectiles might have excited the inclinations and eccentricities of surviving classical KBOs. The scattering is achieved by carefully selecting the number, masses and orbital trajectories of the projectiles. The survival of the resonant KBOs under perturbations from massive projectiles capable of ejecting the Classical KBOs seems unlikely.

4.5   What role have collisions played in shaping the properties of the belt?

The widely held belief that the Kuiper Belt is a collisionally evolved system is based on large extrapolations of the measured KBO population. In fact, the observed KBOs have mutual collision times that are vastly in excess of the age of the solar system: they constitute a collisionless system! What is the present day collision rate in the Kuiper Belt? The answer depends on a careful assessment of the small body population since, in a $q \approx -4$ distribution, the cross-section lies with the smallest objects. The Taiwan-America Occultation Survey is uniquely sensitive to small KBOs and, hopefully, will soon throw light on this question (see http://taos.asiaa.sinica.edu.tw/).

4.6   What are the albedos?

Other than ISO measurements of low ($\sim 3\sigma$) statistical significance, there are no detections of thermal radiation from KBOs. Knowledge of KBO sizes and albedos is therefore confined to measurement of the product of albedo with the geometric cross-section, from optical data alone. Measurements of related objects (nuclei of short-period comets, Centaurs) support the assumption of generally low visual albedos near 4%. It is possible, however, that some (many?) KBOs have much higher albedos. Indeed, the largest KBO (Pluto) retains surface frosts that convey an albedo near 60%. Do reflective frosts coat the surfaces of other KBOs?

4.7   What is the radial extent?

The Kuiper Belt has an apparent edge near 50 AU, in the sense that KBOs have not yet been discovered at distances > 50 AU. Whether this is also a physical edge is not clear. Ultra-deep, wide-field imaging surveys are needed. Theory provides no useful guide to the expected radial extent of the Kuiper Belt.

4.8   What is the ultra-red material?

Ultra-red matter is present on some KBOs but absent from the inner solar system. Is this matter thermally unstable, or buried (or otherwise hidden from view) by fall-back debris ejected in response to outgassing when near the sun (e.g. in the case of the Centaurs)? What is the chemical composition of the ultra-red material and how was it formed?

4.9   Is the Kuiper Belt a local analog of dusty disks around other main-sequence stars?

If the Kuiper Belt were once 100 times more massive than at present, was the cometary flux also 100 times greater and was the dust production rate (which scales roughly with the square of the number of colliding particles) fully 10^4 times greater than at present? Are Kuiper Belts produced naturally as products of accretion in circumstellar disks? What is the time dependence of the dust cross-section in our own Kuiper Belt, and in the Kuiper Belts of other stars?

5   Acknowledgement

The author's work on the Kuiper Belt is conducted in collaboration with Jane Luu and Chadwick Trujillo and is supported by grants from NASA.

Physical Characteristics of Trans-Neptunian Objects and Centaurs

John K. Davies

Joint Astronomy Centre, 660 N A'ohoku Pl., Hilo, Hawaii, 96720, USA.
e-mail: J.Davies@jach.hawaii.edu

Abstract. The available knowledge of the sizes, rotation periods, lightcurve amplitudes, colours and spectra of the Centaurs and Trans-Neptunian Objects is summarised. There are a wide range of reflectances within the Centaur population and no obvious correlations with heliocentric distance. Spectroscopic evidence, including the detection of water ice, points to the Centaurs being large, inactive cometary nuclei. Published photometry of the Trans-Neptunian Objects is often inconsistent, but some conclusions are presented on the issues of spectral diversity within the Kuiper Belt and on the relationship between the Centaurs and the Trans-Neptunian Objects.

1 Introduction

Although no formal definition exists, the Centaurs may be regarded as a group of minor planets following unstable orbits with semi-major axes between those of Jupiter (5.2 AU) and Neptune (30 AU). Eight such objects are presently known. (20 Centaurs are recognised as of July 2000 - Eds). Canonically Centaurs have been identified as asteroids at discovery although (2060) Chiron was subsequently shown to have cometary activity. Jewitt and Kalas [38] define Centaurs as having both perihelia and semi-major axes between Jupiter and Neptune by which definition the comets P/Oterma and P/Schwassmann-Wachmann 1 also qualify for inclusion. The Centaurs are believed to be in the process of diffusing inwards from dynamically chaotic regions within the large reservoir of planetesimals lying beyond Neptune [61,53]. See also the earlier review by Stern and Campins [66] and the discussion of the dynamical fate of Pholus-like objects by Asher and Steele [2].

The Trans-Neptunian Objects, also known as Kuiper Belt Objects after Kuiper [3] and Edgeworth-Kuiper Objects [1] [1], comprise three dynamical classes of more distant objects close to or beyond the orbit of Neptune. Objects in the 3:2 mean motion resonance with Neptune have been described as "Pluto Type" [55] or more commonly "Plutinos" [39], those beyond about 41 AU as "Cubewanos" [56] or "Classical Kuiper Belt Objects", and those

---

[1] For this review I have adopted the term Trans-Neptunian Object (TNO) in preference to KBO or EKO as this was the title proposed by the workshop Scientific Organising Committee and it is the least controversial of the possibilities.

having a much larger semi-major axis and higher eccentricity than the Plutinos and Classical Kuiper Belt Objects are known as Scattered Kuiper Belt Objects [52]. There is to date a single member known of this latter class (1996 TL$_{66}$) although it may be the harbinger of a much larger and more significant population. (26 SKBOs are known as of July 2000 - Eds).

2   Sizes

Since they are not resolved and their albedoes are generally unknown, sizes of these distant objects are best estimated via thermal models similar to those used for asteroids. These require simultaneous, or at least quasi-simultaneous, observations in both the visible and thermal (infrared or mm) regimes.

Such observations have been done for the Centaurs Chiron [46,67,15,1], Pholus [36,20,22], and 1997 CU$_{26}$ [38]. In addition, Bus et al. [14] used an occultation technique to place lower limits on the size of (2060) Chiron and these are in good agreement with the other results. The albedo deduced for (2060) Chiron is typically higher than for (5145) Pholus and 1997 CU$_{26}$, possibly because of resurfacing related to its cometary activity. However, since low albedoes are typical of short period cometary nuclei it has become conventional to assume an albedo of 0.04 for the Centaurs, leading to the remaining estimated diameters quoted in Table 1, which is modified from that presented by Jewitt and Kalas [38].

Table 1. Orbital parameters and sizes for Centaurs

Name	q(AU)	Q(AU)	e	i (deg)	Diameter (km)	Albedo
(2060) Chiron	8.5	18.8	0.38	7	180±10	0.15±0.05
(5145) Pholus	8.7	31.8	0.57	25	190±22	0.04±0.03
(7066) Nessus	11.8	37.4	0.52	16	~78	-
1994 TA	11.7	22	0.3	5	~22	-
1996 DW$_2$	18.9	31	0.24	4	~70	-
(8405) 1995 GO	6.9	29.3	0.62	18	~74	-
1997 CU$_{26}$	13.1	18.4	0.17	23	302±30	0.045±0.010
1998 SG$_{35}$	6.8	11.0	0.23	12	~18	-

Based on the number of detections in their medium depth ecliptic survey, Jewitt et al. [41] suggested that there were of order 2600 Centaurs with diameters greater than 75 km. Jedicke and Herron [37] used the number of Centaurs detected by the Spacewatch automated search programme to estimate that there must be fewer than ~2000 Centaurs, with only three of these having diameters greater than about 200 km. With the discovery of

(2060) Chiron, (5145) Pholus, and 1997 CU_{26}, all of which are close to or over this limit, it may be that these three have already been found. (Sheppard et al. (2000), Astronomical Journal, in press, use a more complete analysis to infer 100 Centaurs (radius r > 50 km) and 10^7 (r > 1 km) -Eds).

Attempts to determine the sizes and albedos of TNOs using the Infrared Space Observatory ISO have only been partly successful [72] but these observations tend to confirm the generally low albedos assumed for these objects.

3   Rotation Periods and Lightcurve Amplitudes

Knowledge of the synodic period of (2060) Chiron has been steadily improved by the work of *e.g.* Bus et al. [13], Luu and Jewitt [47], Dahlgren et al. [18] and Marcialis and Buratti [54]. Luu and Jewitt [47] and Meech and Belton [60] explained the systematic changes in the amplitude of the lightcurve by the diluting effect of Chiron's coma.

Buie and Bus [11] presented a lightcurve of (5145) Pholus and demonstrated convincingly a 9.9825±0.004 hour rotation period with an amplitude of ~0.15 mag; Hoffmann et al. [11] deduced a similar period and amplitude. Davies et al. [24] used the lightcurve parameters of Buie and Bus to fit 13 V magnitudes recorded over 3 days in 1998 and confirmed the 1992 result. Brown and Luu [10] presented a complete lightcurve of (8405) 1995 GO and derived a period of 8.87±0.02 hours and an amplitude ~0.34 mag. Davies et al. [24] combined these data with values from Romanishin et al. [64] and new data from 1997 to argue for a slightly longer period and larger amplitude (~0.55 mag). However, the quoted amplitude differences may reflect different philosophies, the Brown and Luu value being the amplitude produced by minimising the χ^2 value of a sine fit and the Davies et al. value being the actual peak to peak range of a curve which is not truly sinosoidal. W. Brown (private communication) reports that R band photometry of 1997 CU_{26} shows no variation within the likely errors over a period of 5 days, a result consistent with the limited visible photometry of Davies et al. [24] and the infrared measurements of McBride et al. [58].

To date few TNO lightcurves have been published. Using the relatively small number of observations made at the time of its discovery, Williams et al. [75] suggested a possible lightcurve amplitude of 0.5 mag and period of 15 hours for 1993 SC. However, more comprehensive studies by Davies et al. [23] and Tegler et al. [68] placed upper limits of 0.1 and 0.06 respectively on this object's lightcurve and were unable to identify a rotation period. Luu and Jewitt [51] reported that 1996 TL_{66} shows no evidence for rotational modulation greater than 0.05 mag over a period of 6 hours. Similar results come from Delahodde et al. [26] and Collander-Brown et al. [16] which are consistent with the remark of Jewitt and Luu [40] that "rotational modulation of the brightness is typically less than a few tenths of a magnitude although individual objects may of course exceed this". Romanishin and Tegler [65]

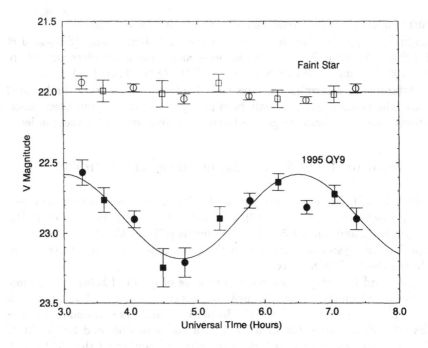

Fig. 1. Lightcurve of 1995 QY$_9$ from Romanishin and Tegler [65].

reported the absence of a detectable lightcurve in a number of the larger TNOs but suggested that 1994 TB, 1994 VK$_8$ and 1995 QY$_9$ have rotation rates of order 6-10 hours and amplitudes of 0.3-0.5 mag. (See Table 2 and Fig. 1). By assuming that the lightcurve variations in the TNOs were due to irregular shapes, and then using the presumed strength of likely icy materials to determine the minimum size of objects likely to collapse into a sphere under self-gravity, Romanishin and Tegler [65] went on to argue for TNOs having albedos close the range of 0.04 found for the Centaurs rather than the higher albedos of bodies such as Pluto and Triton.

4   Colours

4.1   Visible colours

Visible (BVRI) colours for the Centaurs have been reported by Mueller et al. [62], Luu and Jewitt [49], Green et al. [31], Romanishin et al. [64], Brown and Luu [10], Tegler and Romanishin [69], Magnusson et al. [57] and Davies et al. [24]. In general these are in good agreement but there are two exceptions. Brown and Luu determined a V−R color of 0.73±0.04 for (8405) 1995 GO which is significantly different from the value of 0.47±0.04 obtained the same month by Romanishin et al. [64] and from the value of 0.41±0.02 found by

Table 2. Rotation periods and amplitudes

Name	Period (hrs)	Ampl. (mag)	Reference
(2060) Chiron	5.9180 ± 0.0001	0.088 ±0.003	Bus et al. [13]
(2060) Chiron	5.91780 ± 0.00005	0.09-0.045	Luu and Jewitt [47]
(2060) Chiron	∼5.92	0.07 ± 0.02	Dahlgren et al [18]
(2060) Chiron	5.917813 ± 0.000007	0.04	Marcialis and Buratti [54]
(5145) Pholus	9.9825 ± 0.004	∼0.15	Buie and Bus [11]
(5145) Pholus	9.9768 ± 0.001	∼ 0.2	Hoffmann et al. [35]
(8405) 1995 GO	8.87 ± 0.02	0.34	Brown and Luu [10]
(8405) 1995 GO	8.9351 ± 0.0003	∼0.55	Davies et al. [24]
1997 CU$_{26}$	?	Small	Brown (Pers Comm)
1998 SG$_{35}$	>12hr	>0.4	Green (Preliminary result)
1993 SC	?	< 0.1	Davies et al. [23]
1993 SC	?	< 0.06	Tegler et al. [68]
1994 TB	∼ 6	∼ 0.3	Romanishin and Tegler [65]
1996 TL$_{66}$	?	< 0.1	Luu and Jewitt [51]
1996 TO$_{66}$	?	< 0.1	Romanishin and Tegler [65]
1996 TO$_{66}$	∼6.25	∼0.1	Delahodde et al. [26]
1996 TP$_{66}$	?	< 0.07	Collander-Brown et al. [16]
1994 VK$_8$	∼ 10	∼0.4	Romanishin and Tegler [65]
1994 VK$_8$	?	0.6	Collander-Brown et al. [16]
1995 QY$_9$	∼ 8	∼0.6	Romanishin and Tegler [65]

Davies et al. [24] a year later. For 1995 DW$_2$ Luu and Jewitt [49], Green et al. [31], Tegler and Romanishin [69] and Magnusson et al. [57] all give rather different BVRI colours which are not consistent within their quoted errors. In the case of 1995 GO the redder values of V-R seem incompatible with the visible-infrared colours reported elsewhere and so I have adopted the lower values. In the case of 1995 DW$_2$, which is presently much fainter than (8405) 1995 GO, an average value is used. The representative mean colours are given in Table 3.

To date, there has been no evidence of colour variations with rotational phase in any of the well observed Centaurs which eliminates surface heterogeneity on hemispherical scales as a likely explanation for the observed differences in colours reported by different observers.

4.2 Visible-infrared colours

Visible-Infrared, basically V-JHK, colours were shown by Hartmann et al. [32] to be useful in probing the surface compositions of minor planets and comets.

Ideally, to derive visible-IR colours, data over the whole range of wavelengths would be obtained quasi-simultaneously as was done for (2060) Chiron in 1988 [33]. However in reality visible and infrared observations tend to be significantly non-simultaneous and most attempts to derive visible-IR colors for Centaurs (e.g. Hartmann et al. [32], Davies et al. [19,21,22,24] Weintraub et al. [73]) have used V magnitudes estimated using the object's absolute magnitude H_V and assumed values of the phase curve slope parameter G. A fundamental weakness of this approach is that it cannot take account of lightcurve variations unless the lightcurve is known with high precision and the interval between the lightcurve determination and the photometric observations is short enough that the uncertainties in the period cannot build up significantly. These effects may be negligible in the case of the larger Trans-Neptunian Objects, for which no significant lightcurves have yet been detected, but must be considered when dealing with smaller, and potentially irregular objects. In these cases it may be uncertainties in rotational phase, rather than photometric precision, which dominate the uncertainty in the colours.

Combining all the data produces the visible-IR colours given in Table 3 and the reflectivity curves plotted in figure 2. The unusual 'Cometary asteroid' 1996 PW is included in Table 3 for comparison. Note that while Davies et al. [25] and Hicks et al. [34] argue on colour grounds that this is a quiescent comet nucleus, Weissman et al. [74] suggest that it is an asteroid like object returning to the inner solar system after a soujourn in the Oort cloud.

Table 3. Visible–Infrared Colours of Centaurs

Object	B–V	V–R	V–I	V–J	V–H	V–K
(2060) Chiron		0.37±0.03	0.72±0.04	0.9–1.5	1.2–1.8	1.4–1.8
(5145) Pholus	1.27±0.1	0.78±0.03	1.55±0.02	2.59±0.02	2.96±0.02	2.93±0.02
(8405) 1995 GO	0.75±0.04	0.44±0.03	0.96±0.03	1.65±0.02	2.02±0.06	2.11±0.05
(7066) Nessus	1.09±0.04	0.79±0.04	1.50±0.10	2.29±0.04	2.57±0.1	2.57±0.1
1997 CU$_{26}$	0.77±0.05	0.47±0.02	1.01±0.02	1.74±0.02	2.15±0.02	2.22±0.03
1995 DW$_2$	0.54±0.1	0.43±0.07	0.92±0.1	1.31±0.1	–	–
1996 PW	–	0.56±0.04	1.03±0.06	1.80±0.05	2.19±0.05	2.23±0.05

Notes: (2060) Chiron: Variable; (8405) 1995 GO: V–R of 0.73 not used; 1996 PW: Comet-like orbit.

4.3   TNO visible colours

Luu and Jewitt [49], Green et al. [31], Tegler and Romanishin [68], Jewitt and Luu [40] and Magnusson et al. [57] present visible colours which show a wide

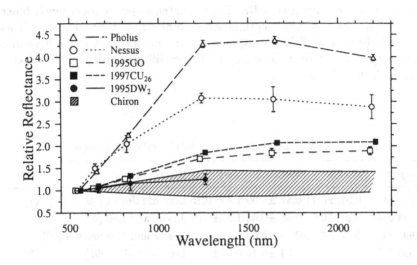

Fig. 2. Reflectivity curves for Centaurs as given by McBride et al. [58].

range of values amongst the TNO population and they argue for significant colour diversity in the TNO population. Barucci et al. [3] confirm this result. However, Tegler and Romanishin [70] suggest on the basis of BVR colours, which they believe are more accurate than those of earlier workers, that these distant objects comprise two populations, with one group of objects being only slightly redder than the Sun (Chiron like) while the others are very red indeed (Pholus like). That such differences in photometry can exist is clear when comparing the BVRI data from several groups as is done in Table 4. In some cases, for example 1996 TP$_{66}$ and 1996 TO$_{66}$, the V−R colours are in good agreement, in others both the V−R, and particularly the B−V, are not consistent within the quoted photometric errors. This phenomenon was also pointed out by Tholen [71] and Barucci et al. [3] who showed that the level of inconsistency far exceeds that expected from the errors quoted by the different groups.

Several factors may be contributing to this effect. Davies et al. [23] emphasized the difficulties involved in photometry of faint moving objects and show, based on reduction of 33 frames of the Kuiper Belt object 1993 SC using different techniques, that the statistical uncertainties of most photometric software packages provide an underestimate of the true observational errors, indeed they obtained significantly different values for the V magnitude of 1993 SC using different reduction techniques on the same frames. It may be significant that almost all of the groups cited in Table 4 appear to have used different filter sets including Mould BVRI, Harris BVR, Kitt Peak VRI and Bessell BVRI. Interestingly the B-V colours of the best observed TNOs derived using the University of Hawaii 88 inch telescope are consistently bluer (ie B-V is numerically smaller) than those obtained at other telescopes. This may be indicative of different responses of various filter/CCD combinations which may make intercomparison of data from different groups more difficult than might first be assumed.

Table 4. Visible colours of TNOs by different groups

Object	B−V	V−R	V−I	Reference	Notes
1993 SC	0.92 ±0.11	0.57 ±0.09	1.43±0.15	Luu and Jewitt [49]	(1)
1993 SC	0.94 ±0.05	0.68 ±0.04	1.36±0.04	Jewitt and Luu [40]	
1993 SC	1.27 ±0.11	0.70 ±0.04		Tegler and Romanishin [70]	
1993 SC		0.54 ±0.14	0.97±0.14	Davies et al. (1997)	(2)
1994 TB	1.10 ±0.08	0.74 ±0.11		Barucci et al. [3]	
1994 TB	0.88 ±0.15	0.85 ±0.15	1.5±0.25	Luu and Jewitt [49]	
1994 TB	1.10 ±0.15	0.68 ±0.06		Tegler and Romanishin [70]	
1994 TB		0.69 ±0.07	1.46±0.10	McBride (*Priv Comm*)	
1996 TL$_{66}$	0.74 ±0.08	0.46 ±0.05	0.72±0.07	Barucci et al. [3]	
1996 TL$_{66}$	0.58 ±0.05	0.13 ±0.06	0.67±0.05	Jewitt and Luu [40]	
1996 TL$_{66}$	0.75 ±0.02	0.35 ±0.01		Tegler and Romanishin [70]	
1996 TL$_{66}$		0.36 ±0.04		McBride (*Priv Comm*)	
1996 TO$_{66}$	0.72 ±0.07	0.34 ±0.04	0.76±0.04	Barucci et al. [3]	
1996 TO$_{66}$	0.59 ±0.04	0.32 ±0.06	0.68±0.05	Jewitt and Luu [40]	
1996 TO$_{66}$	0.74 ±0.03	0.38 ±0.03		Tegler and Romanishin [70]	
1996 TO$_{66}$		0.39 ±0.03	0.73±0.03	McBride (*Priv Comm*)	
1996 TO$_{66}$	0.58 ±0.06	0.42 ±0.07		Delahodde et al. [26]	
1996 TP$_{66}$	1.14 ±0.14	0.65 ±0.13	1.08±0.14	Barucci et al. [3]	
1996 TP$_{66}$	0.80 ±0.07	0.65 ±0.07	1.36±0.06	Jewitt and Luu [40]	
1996 TP$_{66}$	1.17 ±0.05	0.68 ±0.03		Tegler and Romanishin [70]	
1996 TP$_{66}$		0.68 ±0.04	1.01±0.03	McBride (*Priv Comm*)	

Notes: (1) Colours from spectra; (2) V may be unreliable

4.4   TNO visible-infrared colours

Jewitt and Luu [40] extended the issue of colour diversity by presenting V-JHK colours for 5 TNOs. They found a range of V-JHK similar to that amongst the Centaur population and a strong relationship of V-J to M_R with the redder objects being associated with fainter absolute magnitudes (See Fig. 3). If one assumes a roughly constant albedo amongst objects of different colours, this implies the smaller objects are redder. This colour-size relationship would be consistent with the retention of surface frosts, which are generally blue in colour, by the larger objects. However it is worth noting that even the largest objects in this sample are much smaller than Pluto and Triton which are known to have at least partly icy surfaces. Davies et al. [23] and Davies et al. (in prep) independently confirm several of the V-J values although their failure to detect at J some other objects which ought to have been detectable according to the V-J vs M_R relationship, suggests that this relationship may not hold for all TNOs. Luu and Jewitt [49] and Barucci et al. [3] find no correlation of V-R with H_V in a somewhat larger sample of objects. Jewitt and Luu [40] speculate that the failure to detect such a relationship in their earlier data may be because BVRI does not give a long enough baseline for the effect to become apparent, because the other objects were on average smaller or because the apparent correlation is in fact a fluke.

5   Spectroscopy

5.1   The Centaurs

The visible spectra of the 6 brightest Centaurs are featureless (Binzel [4,5], Lazzarin and Barucci [43]), with slopes ranging from flat (2060) Chiron, 1995 DW_2) to red ((5145) Pholus, (7066) Nessus) as would be expected from their visible colours. Fink et al. [30] and Binzel [4] suggested refractory organic materials (tholins) to explain the steep red slope of Pholus and Hoffmann et al. [35] interpreted the data with more detailed models of tholins plus dark but spectrally neutral material.

Luu et al. [48] reported that (2060) Chiron has a featureless infrared spectrum similar to many C type asteroids, as did M. Brown [6] who obtained a similar result for for (8405) 1995 GO. R. Brown et al. [9] presented a 1.4-2.2μm spectrum of 1997 CU_{26} which shows a red spectral slope and absorptions at 1.52 and 2.03 μm indicative of water ice, but no other identifiable spectral features. A similar result was reported by M.Brown and Koresko [7]. The general shape of the near infrared spectrum of 1997 CU_{26} was confirmed by McBride et al. [58].

Infrared spectroscopy of (5145) Pholus revealed a unique spectral feature at a wavelength of about 2.25 μm (Davies et al. [21]) which Wilson et al. [76] proposed might be explained by a mixture of a specific tholin, HCN polymer plus H_2O and NH_3 ices. Luu et al. [48] obtained higher quality IR spectra of

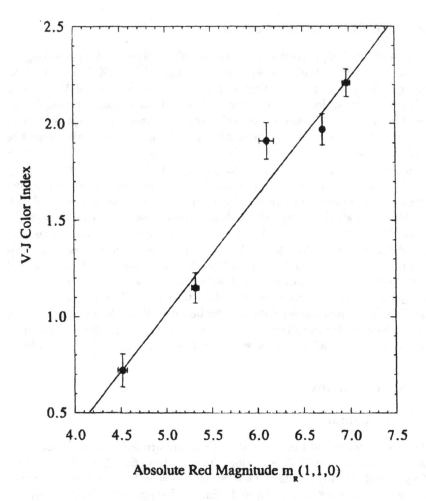

Fig. 3. The V-J vs M_R relationship for 5 TNOs noted by Jewitt and Luu [40].

(5145) Pholus and noted the similarity of the spectrum to a laboratory tar sand. Recently Cruikshank et al. [17] have combined much of the available spectroscopy and photometry of (5145) Pholus and fit this with a model containing a red refractory organic material (tholin), frozen methanol or a photolytic product of that molecule, water ice, olivine and carbon. On this basis they argue that (5145) Pholus is a very primitive object akin to a cometary nucleus which has never been active. The IR spectrum of (5145) Pholus remains unique, observations of a variety of low albedo asteroids such as those in the Cybele, Hilda and Trojans by Luu et al. [48] and Dumas et al. [27] have failed to find an object with a similar 2 μm feature. Noll and Luu [63] and McCarthy et al. [59] report programmes to obtain additional spectroscopic data for Centaurs and TNOs using NICMOS on the Hubble Space Telescope.

5.2   The Trans-Neptunian Objects

The first visible spectrum of a Trans-Neptunian Object was presented by Luu and Jewitt [50] who reported a reddish slope indistinguishable from many asteroids and comet nuclei and noted that the slope was intermediate between those of Centaurs (2060) Chiron and (5145) Pholus. They noted that their spectrum was not well represented by a linear fit and remarked that this might reflect absorption or emission features in the spectrum. Brown et al. [8] presented a 1.4-2.2μm spectrum of 1993 SC which had a strongly red continuum and spectral features similar to those seen on Pluto and Triton. On the basis of these features Brown et al. argued for the presence of simple hydrocarbons such as CH_4, C_2H_6, C_2H_4 or C_2H_2 on the surface of 1993 SC. However the slope of the Brown et al. spectrum is not consistent (being too red) with the H-K colours of Jewitt and Luu [40], a disagreement which is troubling. Luu and Jewitt [51] present an almost continuous spectrum of 1996 TL$_{66}$ between 0.4 and 2.2 μm which is flat and featureless, showing none of the absorption features typical of those of pure ices.

6   Discussion

It does seem clear that the surface colors of the Centaur objects cover a wide range which is representative of objects in the Kuiper Belt. This is consistent with, although does not uniquely demonstrate, that Centaurs are escaped Trans-Neptunian Objects. However is not clear to what extent the properties of the Centaurs can be extrapolated to interpret observations of the Trans-Neptunian Objects. For example, the V-J vs M_R relationship seen in 5 Trans-Neptunian Objects by Jewitt and Luu [40] is not reflected amongst the presently known Centaurs, although this may not be surprising since the Centaurs have already had different evolutionary histories to the TNOs. Weintraub et al. [73] suggested that there might be a trend of decreased reddening with reducing semi-major axis amongst the Centaurs and explained this in terms of increased exposure to the Sun since leaving the Trans-Neptunian region. However this relationship is not supported by the larger dataset of Centaur colours and spectra now available.

The activity of (2060) Chiron, which is not confined to periods around perihelion but occurs sporadically through its entire orbit (Bus et al. [12]) has undoubtedly affected its surface colours. (Note that Chiron was paradoxically quiet close to its recent perihelion passage (eg Lazarro et al. [44,45]). While no such activity has been seen for other Centaurs, given the short observational baselines to date, similar sporadic activity cannot be discounted. It also appears that the standard JHK broad band filters may not be very helpful in distiguishing the spectral characteristics of Centaurs and TNOs since they are too broad to resolve specific absorption bands from water and methane ices. In this respect it can be noted that (8405) 1995 GO and 1997 CU_{26} have almost identical V-JHK colours yet one shows ice absorptions in its spectrum and the other does not.

When presenting their data suggesting bi-modality in the BVR colours, Tegler and Romanishin [70] failed to find any correlation of BVR colour with a number of parameters including semi-major axis, inclination or absolute magnitude and did not advance a specific model to explain their observations. Luu and Jewitt [49] developed a model in which the colour of the TNOs is determined by a balance between cosmic ray induced reddening of a surface dominated by ices and impact induced resurfacing. They suggested that impacts penetrate the irradiation crust and eject material which is less red and less dark than the ancient surface. Since the present observational data refers to colours averaged over a hemisphere, unless one process is dramatically faster than the other this will naturally produce a range of colours. Given the ubiquity of cosmic radiation so far from the Sun and the evidence that large TNOs should have surfaces which have been extensively reworked by impact cratering (Durda et al. [28]) this seems a reasonable explanation of the observations. However at the present time it is impossible to distinguish between a diverse population containing "families" of intrinsically different composition and a weathered or impact gardened population.

In many respects there is an analogy between the present situation of TNO research and that of the early classification of main belt asteroids into a small number of groups based on UBV observations. Only when considerable extra data were available for a larger number of objects was the true structure and complexity of the main belt understood. The same is almost certainly true of the Trans-Neptunian region.

7   Acknowledgements

I extend my thanks to those at both the MBOSS-98 and Flagstaff workshops for sharing their ideas on the interpretation of these data with me and to the MBOSS conference organisers for their invitation to present this review and for their partial financial support. I have attempted to be comprehensive, but I apologise to any friends and colleagues in this field whose publications I may have inadvertantly neglected to cite here and commend to you the excellent EKO web-site maintained by Joel Parker at
http://www.boulder.swri.edu/ekonews/.

References

1. Altenhoff, W.J. and Stumpff, P. (1995): A & A **293** L41-42
2. Asher, D.J and Steele, D.I. (1993): Orbital Evolution of the Large Outer Solar System Object (5145) Pholus. MNRAS **263** 179-190
3. Barucci, M.A., Tholen, D., Doressoundiram, A., Fulchignoni, M. and Lazzarin, M. (1999): Spectrophotometric observations of Edgeworth-Kuiper Belt Objects. These proceedings, BAAS **30** (3) 1112 (abstract) and a paper submitted to Icarus.
4. Binzel, R.P. (1992): The optical spectrum of (5145) Pholus. Icarus **99**, 238-240
5. Binzel, R.P. (1997): 1997 CU_{26}. IAU Circular 6568
6. Brown, M.E. (1998): Infrared Spectroscopy of Centaurs and Irregular Satellites. Presented at the workshop "Exploring the Kuiper Belt". Lowell Observatory. 3-4 September 1998 and BAAS **30** (3) 1112 (abstract)
7. Brown, M.E., Koresko, C. (1998): Detection of Water Ice on the Centaur 1997 CU_{26} ApJ Lett (in Press)
8. Brown, R.H, Cruikshank, D.P., Pendleton, Y., Veeder, G.J. (1997): Surface composition of Kuiper Belt object 1993 SC. Science **276** 937-939
9. Brown, R.H, Cruikshank, D.P., Pendleton, Y., Veeder, G.J. (1998): Identification of Water Ice on the Centaur 1997 CU_{26}. Science **280** 1430-1432
10. Brown, W.R. and Luu, J.X. (1997): CCD photometry of the Centaur 1995 GO. Icarus **126**, 218-224
11. Buie, M.W. and Bus, S.J. (1992): Physical observations of (5145) Pholus. Icarus **100**, 288-294
12. Bus, S.J., A'Hearn, M A, Bowell, M.J. and Stern, S.A. (1998): 2060 Chiron: Evidence for Activity Near Aphelion. Icarus In Press.
13. Bus, S.J., Bowell, E., Harris, A.W. and Hewitt, A.V. (1989): 2060 Chiron: CCD and electronographic photometry. Icarus **77**, 223-238
14. Bus, S.J. et al. (20 Authors) (1996): Stellar Occultation by 2060 Chiron. Icarus **123** 478-490
15. Campins H., Telesco, C.M., Osip, D.J. Rieke, G.H., Rieke M.J. and Schulz, B. (1994): The color temperature of 2060 Chiron — a warm and small nucleus. AJ **108**, 2318-2322
16. Collander-Brown, S.J., Fitzsimmons, A., Fletcher, E., Irwin, M.J., Williams, I.P. (1999): The light curves of the Tran-Neptunian Objects 1996 TP_{66} and 1994 VK_8. These proceedings
17. Cruikshank et al. (15 Authors) (1998): The Composition of Centaur 5145 Pholus. Icarus **135** 389-407
18. Dahlgren, M., Fitzsimmons, A., Lagerkvist, C-I., and Williams, I.P. (1991): Differential CCD Photometry of Dubiago, Chiron and Hektor. MNRAS **250** 115-118.
19. Davies J.K., and Sykes, M.V. (1992): 1992 AD. IAU Circular 5480
20. Davies, J.K., Spencer, J., Sykes, M., Tholen, D. (1993a): IAU Circ 5698. 27 January 1993
21. Davies, J.K., Sykes M.V. and Cruikshank, D.P. (1993b): Near infrared photometry and spectroscopy of the unusual minor planet (5145) Pholus (1992AD). Icarus **102**, 166-169
22. Davies, J.K., Tholen, D.J. and Ballantyne, D.R. (1996): Infrared observations of distant asteroids. In *Completing the Inventory of the Solar System*, ASP

Conference Proceedings (Eds. T.W. Rettig and J.M. Hahn), **107**, pp. 97–105, ASP, San Francisco

23. Davies, J.K., McBride, N. and Green, S.F. (1997): Optical and infrared photometry of Kuiper Belt object 1993 SC. Icarus **125**, 61-66

24. Davies, J.K. McBride, N, Ellison, S.E. Green, S.F. and Ballantyne D. (1998a): Visible and Infrared Observations of Six Centaurs. Icarus **134** 213-227

25. Davies, J. K., McBride, N. Green, S.F., Mottola, S. Carsenty, U. Basran, D. Hudson, K.A. and Foster M.J. (1998b): The lightcurve and colors of unusual minor planet 1996 PW. Icarus **132** 418-430

26. Delahodde, C., Hainaut, O., Boehnhardt, H., Dotto, E., Barucci, E., West, R., Meech, K. (2000): Physical Observations of 1996 TO$_{66}$. These proceedings.

27. Dumas, C. Owen, T. and Barucci, M.A. (1998): Near Infrared Spectroscopy of Low Albedo Surfaces of the Solar System: Search for the Spectral Signature of Dark Material. Icarus **133** 221-232

28. Durda, D and Stern, S.A. (1998): Collisional and Cratering Rates in the Kuiper Belt. Presented at this Workshop.

29. Edgeworth, K.E. (1943): The evolution of our planetary system. J. Brit. Astron. Assoc. **53**, 181-188

30. Fink, U., Hoffmann, M., Grundy, W., Hicks, M. and Sears, W. (1992): The steep red spectrum of 1992AD: an asteroid covered with organic material? Icarus **97**, 145–149

31. Green, S.F., McBride, N. O'Cealleagh, D., Fitzsimmons, A. Irwin M.J. and Williams I.P. (1997): Surface reflectance properties of distant solar system bodies. MNRAS **290**, 186–192

32. Hartmann, W.K., Cruikshank, D.P. and Degewij, J. (1982): Remote comets and related bodies: VJHK colorimetry and surface materials. Icarus **52**, 377–408

33. Hartmann, W.K., Tholen, D.J., Meech K.J. and Cruikshank, D.P. (1990): 2060 Chiron: colorimetry and cometary behavior. Icarus **83**, 1–15

34. Hicks, M.D., Buratti, B.J., Newburn, R.L., Rabinowitz, D.L. (1999): Physical Observations of 1996 PW and 1997 SE$_5$. Extinct Comets or D-type asteroids. submitted to Icarus.

35. Hoffmann, M., Fink, U., Grundy, W.M. Hicks, M. (1993): Photometric and Spectroscopic Observations of (5145) Pholus. J. Geophys. Res. **98 E4** 7403-7407

36. Howell, E., Marcialis, R. Cutri, R. Nolan, M., Lebofsky, L. and Sykes, M. (1992): IAU Circ 5449 12 February 1992

37. Jedicke R.J. and Herron, J.D. (1997): Observational constraints on the Centaur population. Icarus **127**, 494–507

38. Jewitt, D and Kalas, P. (1998): Thermal Observations of Centuar 1997 CU$_{26}$ ApJ **499** L103-106

39. Jewitt, D. and Luu, J.X. (1996): The Plutinos. In *Completing the Inventory of the Solar System*, ASP Conference Proceedings (T.W. Rettig and J.M. Hahn, Eds.), **107**, pp. 255–258, ASP, San Francisco

40. Jewitt, D. and Luu, J.X. (1998): Optical Infrared Spectral Diversity in the Kuiper Belt. AJ **115** 1667-1670

41. Jewitt, D., Luu, J. and Chen, J. (1996): The Mauna Kea-Cerro-Tololo Kuiper Belt and Centaur Survey. AJ **112** 1225-1238

42. Kuiper, G. P. (1951): In *Astrophysics* ed J. A.Hynek. McGraw Hill, New York 357

43. Lazzarin, M. and Barruci M.A. (1998): The Centaurs. Poster presented at the workshop "Exploring the Kuiper Belt" Lowell Observatory. 3-4 September 1998, presented at this Workshop BAAS **30** (3) 1114 (abstract)

44. Lazzaro, D., Florczak, M.A., Angeli, C., Carvano, J.M., Betzler, A.S., Casati, A .A., Barucci, M.A., Doressoundiram, A. and Lazzarin, M.(1997): Photometric Monitoring of 2060 Chiron's brightness at Perihelion. Planet. Space Sci. 45, 1607-1614

45. Lazzaro, D., Florczak, M.A., Betzler, A., Winter, O.C., Giuliatti-Winter, S.M. Angeli C.A., and Foryta, D.W. (1996): 2060 Chiron back to a minimum of brightness. Planet and Space Sci. **44**, 1547-1550

46. Lebofsky, L.A., Tholen, D.J., Rieke G.H. and Lebofsky, M.J. (1984). 2060 Chiron: Visual and thermal infrared observations. Icarus **60**, 532-537

47. Luu, J.X. and Jewitt, D. (1990): Cometary activity in 2060 Chiron. AJ **100** 913-932

48. Luu, J.X., Jewitt D.,and Cloutis, E. (1994): Near infrared spectroscopy of primitive solar system objects. Icarus **109**, 133-144

49. Luu, J.X. and Jewitt, D. (1996a): Color diversity among the Centaurs and Kuiper Belt objects. AJ **112** (5), 2310-2318

50. Luu, J.X. and Jewitt, D. (1996b): Reflectance Spectrum of the Kuiper belt Object 1993 SC. AJ **111** (1), 449-503

51. Luu, J.X. and Jewitt, D. (1998): Optical and Infrared Reflectance Spectrum of Kuiper Belt Object 1996 TL$_{66}$. ApJ **494** L117-121

52. Luu J.X., Marsden, B., Jewitt, D.,Trujillo, C., Hergenrother, C., Chen, J. and Offutt, W. (1997): Nature **387**, 573

53. Malhotra R., (1996): A.J. **111**, 504-516

54. Marcialis, R.L. and Buratti, B.J. (1993): CCD photometry of 2060 Chiron in 1985 and 1991. Icarus **104** 234-243

55. Marsden B. (1996): Searches for Planets and Comets. In *Completing the Inventory of the Solar System*, ASP Conference Proceedings (Eds. T.W. Rettig and J.M. Hahn.), **107**, pp. 193-207, ASP, San Francisco

56. Marsden B. (1997): M.P.E.C. 1997-P12

57. Magnusson, P., Lagerkvist, C-I., Lagerros, J.S.V., Dahlgren, M., and Lundstrom, M. (1998): Observations of Distant Solar System Bodies. Astron. Nachr **319** (4) 251-255

58. McBride, N., Davies, J.K., Green, S.F. and Foster, M.J. (1999): Optical and infrared observations of the Centaur 1997 CU$_{26}$. Submitted to MNRAS October 1998

59. McCarthy, D.W., Campins, H., Kern, S. Brown, R.H., Stolovy, S., and Rieke, M. (1998): 1-2 Micron Spectroscopy of Centaurs and Kuiper Belt Objects from NICMOS BAAS **30** (3) 1114 (abstract)

60. Meech, K.J. and Belton, M.J.S.(1990): The atmosphere of 2060 Chiron. AJ **100**, 1323-1338

61. Morbidelli A., Thomas F., Moons M., (1995): Icarus, 118, 322-340

62. Mueller, B.E.A., Tholen, D.J., Hartmann, W.K., and Cruikshank, D. (1992): Extraordinary colors of asteroidal object (5145 Pholus) 1992AD. Icarus **97**, 150-154

63. Noll, K. and Luu, J.X. Physical Studies of Kuiper Belt Objects with NICMOS. BAAS **30** (3) 1112 (abstract)

64. Romanishin, W., Tegler, S.C., Levine, J. and Butler, N. (1997): BVR photometry of Centaur objects 1995 GO, 1993 HA$_2$ and (5145) Pholus. Astron. J. **113**, 1893–1898

65. Romanishin, W. and Tegler, S.C. (1998): Sizes and shapes of Kuiper Belt Objects Presented at the workshop "Exploring the Kuiper Belt" Lowell Observatory. 3-4 September 1998.

66. Stern A. and Campins, H. (1996): Chiron and the Centuars: escapees from the Kuiper belt. Nature **382**, 507–510

67. Sykes, M.V and Walker, R.G. (1991): Science **251** 777-780

68. Tegler, S.C. and Romanishin, W. (1997): The extraordinary colors of trans-Neptunian objects 1994 TB and 1993 SC. Icarus **126**, 212–217

69. Tegler, S.C., Romanishin, W., Stone, A., Tryka, K., Fink, U. and Fevig, R. (1997): Photometry of the Trans Neptunian Object 1993 SC. AJ 114 1230-1233

70. Tegler, S.C. and Romanishin, W. (1998): Two Distinct Populations of Kuiper Belt Objects. *Nature* **392** (6671) 49-51

71. Tholen, D.J. (1998): Practical Considerations for Photometry of Faint Solar System Objects. Presented at the workshop "Exploring the Kuiper Belt" Lowell Observatory. 3-4 September 1998

72. Thomas, N., Eggers, S., Ip, W.-H., Lichtenberg, G., Fitzsimmons, A., Keller, H.U., Williams, I.P., Hahn, G., Rauer, H. (2000). These proceedings.

73. Weintraub, D.A., Tegler S.C., and Romanishin, W. (1997): Visible and near infrared photometry of the Centaur objects 1995 GO and (5145) Pholus. Icarus **128** 456-463

74. Weissman, P.R. and Levison H.F., (1997): Origin and Evolution of the Unusual Object 1996 PW: Asteroids from the Oort Cloud? ApJ **488** L133-136

75. Williams, I.P., O'Ceallaigh, D.P. Fitzsimmons, A. and Marsden, B.G. (1995): The slow moving objects 1993SB and 1993SC, Icarus, **116**, 180–185

76. Wilson, P.D, Sagan, C. and Thompson, W.R. (1994): The organic surface of (5145) Pholus: Constraints set by Scattering Theory. Icarus **107** 288-303

Detection of Thermal Emission from 1993 SC
– First Results

Nicolas Thomas[1], Sönke Eggers[1], Wing-Huen Ip[2], Günter Lichtenberg[1],
Alan Fitzsimmons[3], Horst Uwe Keller[1], Iwan Williams[4], Gerhard Hahn[5],
and Heike Rauer[5]

[1] Max-Planck-Institut für Aeronomie, D-37189 Katlenburg-Lindau, Germany
[2] Institute of Space Science, National Central University, Chung Li, Taiwan.
[3] Dept. of Pure and Applied Physics, The Queen's University, Belfast, N. Ireland.
[4] Queen Mary and Westfield College, University of London, E1 4NS, Great
   Britain
[5] DLR für Planetenerkundung, Berlin-Adlershof, Germany.

Abstract. We report on observations of the Kuiper-Belt Objects, 1993 SC and
1996 TL$_{66}$ using the European Space Agency's Infrared Space Observatory. The
measurements indicate a first detection of 1993 SC at 90 microns and yield a measurement of the radius of 164 ($\pm$35) km and a geometric albedo of 0.022 ($\pm$0.013).
For 1996 TL$_{66}$, an effective radius of 316($\pm$49) km and a geometric albedo of
0.030($\pm$0.015) have been derived although, in this case, it cannot be ruled out
that the signal is not actually from the KBO. This paper represents a brief report
of an article published more fully elsewhere [13].

1 Introduction

Because of their possible association with short-period comets, it is widely
assumed that Kuiper-Belt Objects (KBOs) must have a geometric albedo (p)
of around 0.04 (cf. [6]) and thus, the observed brightnesses can be converted
in the standard way to effective radii, R, in excess of 150 km for the brightest
objects. However, the relationship between KBOs, on the one hand, and
asteroids and comets, on the other, is by no means clear and it should not be
forgotten that the geometric albedo of Pluto, often referred to as the largest
Kuiper Belt Object, is 0.51!

To verify the albedos and radii of KBOs from Earth, simultaneous (or
quasi-simultaneous) observations of the reflected solar flux in the visible and
the thermal emission flux must be made. Jewitt and Kalas [5] have recently
measured the thermal emission at 20 microns from the Centaur, 1997 CU$_{26}$
(currently at 13.6 AU), using the 3.8-m UKIRT telescope on Mauna Kea.
Knowing the visual magnitude, they derived p=0.045 ($\pm$0.010) and R=151
($\pm$15) km. Centaurs are generally thought to be transition objects in the
process of moving from Kuiper Belt type orbits towards short-period comet-
like orbits. The chaotic nature of the orbit of the Centaur, 2060 Chiron [10],
and its observed activity (e.g. [11]) offer some support for this hypothesis.

Because of their extreme distance, the surface temperatures of KBOs are only around 75 K and, thus, the peak of their thermal emission occurs in the range 60 - 100 microns [12], a wavelength at which the Earth's atmosphere is essentially opaque. However, the launch of the European Space Agency's Infrared Space Observatory (ISO) offered the opportunity to perform observations at these wavelengths. We report here on a set of observations of the Kuiper Belt Objects, 1993 SC and 1996 TL$_{66}$ obtained with ISO.

2   ISOPHOT Observations

The ISOPHOT C100 3 x 3 pixel imaging detector was used since the standard photometer was not performing to its pre-launch specifications. Each pixel covered 43.5 x 43.5 arcsec on the sky. The distance from pixel centre to pixel centre was 46 arcsec [8]. The 90 micron filter was selected. In addition to the zodiacal light, there is also background noise from astronomical objects that may be in the field of view necessitating a complex observational strategy.

The field containing the KBO was observed in triangular chopped mode (ISOPHOT observation template P22) using a chopper throw of 150 arcsec. This measurement contained the KBO, astronomical background objects in all chopper positions, and the elongation angle (Sun-ISO-object) dependent contamination of the zodiacal light. A second measurement of the same area on the sky was acquired several days later once the KBO had moved out of the field. The difference in the roll angle of the spacecraft between the background and the on-source measurements was less than 1 degree for all observations and therefore the target and chopper positions were obtained at the exact same positions on the sky. Subtracting the background from the on-source measurement removed the astronomical background, while subsequently subtracting the difference in the chopper levels between background and on-source removed the elongation angle dependent dust contamination.

3   Reduction

The data reduction was carried out using the software package, PIA (Phot Interactive Analysis) [3], following the standard reduction method for most steps. After obtaining the relative counts for all chopper levels, the top and bottom chopper plateaux were averaged and subtracted from the central level to yield a signal that, as a last step, was flux-calibrated, using the first of the two calibration sources (FCS 1) of the telescope. The absolute calibration of ISOPHOT remains somewhat uncertain. As a result, there is a systematic error of about 20% that must be added to the final error budget.

When performing the internal background subtraction between the different chopper levels, the computed signal is normally underestimated by a factor, g, of 2.75 in the triangular chopped mode. At low flux levels, however, this correction factor appears to be much lower. Although the exact value is not known, g can be estimated by comparing the signal levels derived for the two ISOPHOT calibration sources, which are measured in different chopper positions. Hence, their uncorrected fluxes should give some indication of the magnitude of g that would apply to our datasets. In practice, these differ by only a few percent, implying a value for g that is close to 1. We have assumed $g = 1$ in this analysis.

The weighted average signal from 1993 SC, from 5 individual on-source measurements on 1993 SC and 3 backgrounds is 11.46 ($\pm$4.24) mJy. For 1996 TL$_{66}$, the two on-target measurements and two backgrounds yield 39.77 ($\pm$11.62) mJy corresponding to 3.4 sigma. Two 1993 SC measurements were obtained in 1996 and a further 3 in 1997. Looking at the results for each year individually gives 13.95 ($\pm$7.05) mJy (1996 data) and 10.05 ($\pm$5.31) mJy (1997 data set).

The pixels in the ISOPHOT 3 x 3 array are numbered so that pixel 5 is at the centre. We have confirmed that 1993 SC should have been within 4 arcsecs of the centre of the detector array for both sets of observations.

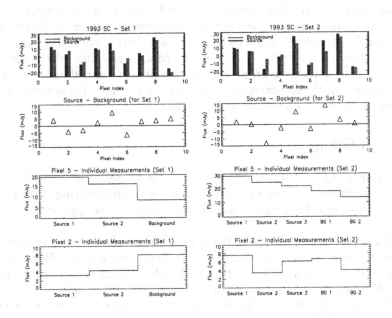

Fig. 1. The individual pixel measurements for observations of 1993 SC. Set 1 corresponds to source measurements made in 1996, set 2 corresponds to source measurements made in 1997. Top: The mean flux in each pixel for source and background. 2nd row: Source minus background for each pixel. Note that pixel 5 has a similar value in both sets. 3rd row: The individual measurements of pixel 5 for all source and background observations. Source measurements are always higher than the background. Bottom row: The individual measurements of pixel 2 for comparison.

In Figure 1, the source and background data for 1993 SC are given for each individual pixel. The data have been split into two sets corresponding to 1996 measurements and 1997 measurements. The four plots on the left are for set 1 (1996 data) and on the right for set 2 (1997 data). The second row from the top shows source minus background for each pixel. It can be seen that in both sets of data, pixel 5 has a similar positive value close to 10 mJy. We note that pixels 3 and 7 show large deviations from zero in set 2. However, pixel 5 has the second highest flux in set 2 and the highest flux in set 1. The probability of pixel 5 having either the highest or second highest value in two independent, random data sets acquired one after the other, is 0.049 or less than 1 in 20. The third row from the top in Figure 1 shows the individual measurements of pixel 5 within each set. In all cases, pixel 5 shows a higher flux when 1993 SC is in the field. The probability of this occurring in a random distribution is just over 3%. The lowermost row shows the behaviour of pixel 2 for comparison. These results indicate a high probability that an object was detected in the central pixel of the C100 detector when ISOPHOT was pointed at 1993 SC.

A similar approach for the 1996 TL_{66} data shows that unlike 1993 SC, the signal is not in pixel 5 but is mostly in pixel 1. Pixel 1 was consistently much higher than the background in the two measurements acquired on target. The observed flux therefore cannot be the result of a short transient (e.g. a glitch). If the detection is of 1996 TL_{66}, then we need to establish why it appears to be in pixel 1 rather than pixel 5. At the present time, despite verifying the accuracy of the ephemeris and checking the ISO pointing history, we are unable to provide an adequate explanation.

4   Modelling

The quality of the data does not warrant using a detailed model such as the one produced by [12] and we restrict ourselves to investigating the simplest cases of a standard thermal model (STM or slow rotator approximation; see [9]).

For a Lambertian surface, the hemispherical albedo, A_H, (sometimes called the directional-hemispherical reflectance) is equal to the Bond albedo (A_B) and $A_H = A_B = pq$ where q is the phase integral and equal to 3/2 [12]. Most planetary surfaces are non-Lambertian and we adopt a value for q of 0.75 (see, for example, [5]) although this assumption has relatively little effect on the final results.

5   Results

The detection of a KBO with ISO was only possible if the object has a low albedo. A high visual albedo (>0.15) for the KBO would mean a relatively small diameter in order to conform to the visual magnitude constraint. This implies a much reduced thermal IR flux. Our results indicate that 1993 SC and possibly 1996 TL_{66} have been detected which immediately implies that the albedos of these KBOs, and hence their surface properties, are different from that of Pluto.

For an STM, the combined observations of 1993 SC yield a mean $R=164$ (± 35) km. The individual data sets are consistent within error bars with these values (Figure 2). 1996 TL_{66} is roughly twice the size of 1993 SC having $R=316(\pm 49)$ km.

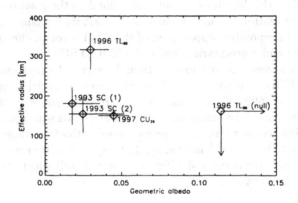

Fig. 2. Albedo and effective radii for 1993 SC and 1996 TL_{66}. For 1993 SC, the results for the two different data sets are shown. For 1996 TL_{66}, the limits for a null detection are also shown.

Figure 2 also shows the albedos. 1993 SC has a mean $p=0.022$ (± 0.013) while for 1996 TL_{66} $p=0.030(\pm 0.015)$ assuming an STM. All are extremely low and comparable to the accepted values for the albedos of cometary nuclei. The albedos for 1993 SC are just consistent with 1997 CU_{26} taking into account the formal error [5]. Davies et al. [1] adopted a visual magnitude for 1993 SC of 22.35 (± 0.10), which would increase the STM estimate for p to 0.028 (± 0.015), in good agreement with the 1996 TL_{66} result.

Although the signal in the ISOPHOT measurement of 1996 TL_{66} is not in the correct pixel, we at present believe that the signal is indeed from 1996 TL_{66}. Although we have no way to prove this, the signal level is consistent with this assumption. If, however, the observation indicates a null detection then the STM indicates a lower limit for p of 1996 TL_{66} of 0.11 and an upper limit for R of 165 km assuming a detection threshold of 10 mJy.

6   Summary and Discussion

Five individual observations of 1993 SC were acquired using the Infrared Space Observatory (ISO) at 90 micron with the ISOPHOT experiment. A 2.7 sigma detection (a confidence level of 99.6%) of the Kuiper Belt Object, 1993 SC, was recorded at a signal level of 11.46 ($\pm$4.24) mJy using ISO. The results have been modelled using a standard thermal model (STM) and give an effective radius of 164 ($\pm$35) km as a lower limit. This corresponds to a geometric albedo of 0.022 ($\pm$0.013).

Two individual observations of 1996 TL_{66} were also obtained. A clear signal of 39.77 ($\pm$11.62) mJy at 90 microns was recorded. However, the position of the signal on the detector does not correspond to the position expected. A detailed investigation has not revealed a satisfactory explanation. Assuming that ISO was mispointed, application of the STM gives an effective radius of 320($\pm$49) km and a geometric albedo of 0.030($\pm$0.015).

This is in agreement with expectations based on the assumption that the surfaces of KBOs are similar to those of cometary nuclei. The results for 1996 TL_{66} and 1993 SC show that KBOs are large and very dark objects. Their surfaces cannot be composed of the surface frosts which produce Pluto's relatively high geometric albedo. The escape velocities from these large bodies are typically 100-250 m s^{-1} indicating that any comet-like emission from the surface [2] must overcome a significant potential well which may affect the dynamics of any outflow.

References

1. Davies, J.K., McBride, N., and Green, S.F. 1997, Icarus, 125, 61-66
2. Fitzsimmons, A., et al., 1998. Presented at this Workshop.
3. Gabriel, C., et al, 1998, "The Isophot interactive analysis -PIA (Version 7.0)", European Space Agency.
4. Gabriel, C., et al, 1997, "Proc. of the ADASS VI Conference", eds. G. Hunt and H.E. Payne, pp. 108.
5. Jewitt, D. and Kalas, P. 1998, ApJ, 499, L103-L106
6. Keller, H.U., Curdt, W., Kramm, J.-R. and Thomas, N. 1996 In "Images of the Nucleus of Comet P/Halley", ESA SP-1127, Vol. 1, Eds. Reinhard, R., Longdon, N. and Battrick, B.
7. Klaas, U., et al., 1994, "Isophot Observers Manual (Version 3.1)", European Space Agency.
8. Laureijs, R. J., et al, 1996, "Isophot Data Users Manual (Version 2.0)", European Space Agency.
9. Lebofsky, L.A. and Spencer, J.R. 1989, Radiometry and thermal modelling of asteroids, in Asteroids II, eds. Binzel, R.P., Gehrels, T., Matthews, M.S., University of Arizona Press, Tucson, 128.
10. Oikawa, S. and Everhart, E. 1979, AJ, 84, 134-139
11. Rauer, H. et al. 1997, Planet. Space Sci, 95, 799-805

12. Thomas, N., Fitzsimmons, A., and Ip, W.-H. 1997, Planet. Space Sci., 45(3), 295-309
13. Thomas, N., Eggers, S., Ip, W.-H., Lichtenberg, G., Fitzsimmons, A., Jorda, L., Keller, H. U., Williams, I. P., Hahn, G., Rauer, H. 2000. Astrophys. J., 534, 446-455

Physical Characteristics of Distant Comets

Dina Prialnik

Department of Geophysics and Planetary Sciences
Tel Aviv University, Ramat Aviv 69978, Israel

Abstract. The initial structure of a comet nucleus is most probably a fine-grained porous material composed of a mixture of ices, predominantly H_2O, and dust. The water ice is presumably amorphous and includes occluded gases. This structure is bound to undergo significant changes during the long residence of the nucleus in the Oort cloud or the Kuiper belt, due to internal radiogenic heating. The evolved structure of a comet nucleus is thus far from homogeneous: the porosity and average pore size change with depth and the composition is likely to become stratified. Such changes occur mainly as a result of gas flow through the porous medium: different volatiles — released by sublimation or crystallization of the amorphous ice — refreeze at different depths, at appropriate temperatures, and the gas pressure that builds up in the interior is capable of breaking the fragile structure and alter the pore sizes and porosity. These processes have been modelled and followed numerically. However, many simplifying assumptions are necessary and the results are found to depend on a large number of uncertain parameters. Thus porous comet nuclei may emerge from the long–term evolution far from the sun in three different configurations, depending on the thermal conductivity, porous structure, radius, etc.: a) preserving their pristine structure throughout; b) almost completely crystallized (except for a relatively thin outer layer) and considerably depleted of volatiles other than water and c) having a crystallized core, layers including large fractions of other ices and an outer layer of unaltered pristine material. Liquid cores may be obtained if the porosity is very low. The extent of such cores and the length of time during which they remain liquid are again determined by initial conditions, as well as by physical properties of the ice. If, in addition to the very low porosity, the effective conductivity is low, it seems possible to have both an extended liquid core, for a considerable period of time, and an outer layer of significant thickness that has retained its original pristine structure.

1 Introduction

Are comets pristine bodies that hold clues to the formation of the solar system? Are comets carriers of prebiotic organic molecules, spreading the seeds of life throughout the solar system by impacts on other bodies? Can comets be both? If so, can one comet be partly pristine and partly evolved or are some comets pristine and others evolved? These are the questions that we shall address in this paper.

The concept that comets are pristine objects, since they spend most of their lives at very large heliocentric distances, unaffected by solar radiation, is entirely based on the assumption that comets are devoid of internal heat sources. But a large fraction of cometary material is (meteoritic) dust and it is therefore reasonable to assume that it includes radioactives. One may then argue that, being small objects, comets have a large surface to volume ratio and hence may get rid of internal heat quite efficiently and remain cold in spite of the radiogenic heat sources. But cometary material is estimated to have a very low thermal conductivity, due both to the nature of the ice, which is presumably amorphous (cf. Mekler and Podolak 1994), and to the grainy porous structure (cf. Greenberg 1998). Consequently, the rate of cooling may not be efficient enough, particularly if the heating power is high, as would be expected, for example, from the short-lived radioactive isotope ^{26}Al. Indeed, early estimates of the potential role of ^{26}Al in the thermal evolution of comets, by Irvine et al. (1980) and Wallis (1980) (see Podolak and Prialnik 1997 for a review) showed that it might be sufficient for melting the ice.

Thus the early evolution of distant comets seems worthy of more detailed investigation, based on numerical simulation; an outline of the method generally employed is given in Sect. 2 below. Some knowledge of the initial, pristine, structure and composition is required, however, as input to these computations. As shown further in Sect. 2, general properties of comet nuclei may be inferred from their activity at large heliocentric distances. Beyond that, one is forced to resort to parameter studies, considering wide ranges of variation of the most relevant properties. These parameters may be identified based on general, analytical, arguments, as shown in Sect. 3. Results of parameter studies are briefly described in Sect. 4 and the main conclusions are summarized in Sect. 5.

2   Modelling the Structure and Evolution of Comets

2.1   The set of evolution equations

Real comet nuclei need not have any geometrical symmetry (self gravity being negligible), but comet nucleus models must assume some form of symmetry, for the sake of simplicity, clarity, and in order to minimize the number of free parameters. Thus sphericity is the common assumption for the nucleus shape. The simplest among such models is a one-dimensional, spherically *symmetric* nucleus. It is particularly suitable for evolution at large heliocentric distances, which is governed by internal processes, rather than by non-homogeneous insolation.

The most general composition of a comet nucleus includes water ice — amorphous and crystalline, water vapor, dust, and other volatiles, which may be frozen, free, or trapped in the amorphous water ice. Let ρ denote the bulk mass density, and the density of the various components be denoted by ρ_a (amorphous ice), ρ_c (crystalline ice), ρ_v (water vapor), ρ_d (dust), $\rho_{s,n}$ and $\rho_{g,n}$, where the index n runs over the different species of volatiles other than H_2O (such as CO, CO_2, etc.), in solid (s) or gaseous (g) form. The amorphous ice includes (small) mass fractions f_n of trapped gases. The implicit assumption of the model is that all the components of the nucleus are in local thermal equilibrium and hence a unique local temperature may be defined. The energy per unit mass u is given by $\rho u = \sum_\alpha \rho_\alpha u_\alpha$, where the specific energies u_α may be functions of the temperature T. Denoting mass fluxes by $\mathbf{J}_\alpha$, the rates of sublimation (condensation) of the volatiles by q_α, and the temperature dependent rate of crystallization of amorphous ice, as given by Schmitt et al.(1989), by $\lambda(T)$, we write the set of equations that describes the evolution of a comet nucleus. The mass balance equations are:

$$\frac{\partial \rho_a}{\partial t} = -\lambda(T)\rho_a, \tag{1}$$

$$\frac{\partial \rho_c}{\partial t} = (1 - \sum_n f_n)\lambda(T)\rho_a - q_v, \tag{2}$$

$$\frac{\partial \rho_v}{\partial t} + \nabla \cdot \mathbf{J}_v = q_v, \tag{3}$$

for H_2O and similar equations for the other volatiles. The evolution of the porosity p (the fraction of empty volume) is obtained, locally, from the changes in density. The energy conservation law is

$$\frac{\partial}{\partial t}(\rho u) + \nabla \cdot (\mathbf{F} + \sum_\alpha u_\alpha \mathbf{J}_\alpha) = -\sum_\alpha q_\alpha \mathcal{H}_\alpha + \lambda(T)\rho_a(1 - \sum_n f_n)\mathcal{H}_{ac} + Q, \tag{4}$$

where $\mathcal{H}$ is the latent heat of sublimation, $\mathcal{H}_{ac}$ is the energy released in crystallization of the amorphous ice and Q is the rate of radiogenic energy release,

$$Q = \rho_d \sum_j \tau_j^{-1} X_{0,j} \exp^{-t/\tau_j} H_j, \tag{5}$$

where τ_j is the characteristic decay time of the $j'th$ radioactive isotope, $X_{0,j}$ – its initial mass fraction within the dust, and H_j – the energy released per unit mass upon decay. The heat flux is $\mathbf{F} = -\psi(p)K(T)\nabla T$, where $K(T)$ represents the thermal conductivity of the mixture of ices and dust, corrected by a factor $\psi(p) < 1$ due to the reduced contact surface between grains in a porous material; $\psi(p)$ may be orders of magnitude smaller than unity (e.g., Steiner & Kömle 1991). The gas flux in a porous medium depends both on the porosity and on the pore size (cf. Mekler et al. 1990). When the pore size is small (i.e., the corresponding Knudsen number exceeds unity), the flow of gas is essentially a free molecular (Knudsen) flow. Pores may grow in size due to the internal gas pressure (Prialnik et al. 1993); they may also grow (or shrink) due to sublimation (recondensation) of volatiles from (onto) their surfaces.

The boundary conditions are vanishing fluxes (mass and heat) at the center of the nucleus and vanishing pressure at the surface R; since at large heliocentric distances d_H there is no surface sublimation, the surface heat flux is given by

$$\mathbf{F}(R,t) = 4\pi R^2 [(1-A)L_\odot/(16\pi d_H(t)^2) - \epsilon\sigma T(R,t)^4], \qquad (6)$$

where A is the albedo, $L_\odot$ is the solar luminosity, ϵ is the emissivity (of order 1) and σ is the Stefan-Boltzmann constant. The system of non-linear time-dependent second order partial differential equations is turned into an implicit difference scheme and solved iteratively (cf. Prialnik 1992, Espinasse et al. 1993, Tancredi et al. 1994, Benkhoff and Huebner 1995).

The procedure of comet modelling is to adopt a set of parameters for the cometary structure, follow the evolution in a given orbit and then adjust this set until agreement is obtained with observations (e.g., regarding production rates and surface temperature). Of course, in principle, such a set may not be unique; but a set of parameters which are compatible with each other and yield a good match to observations may be taken to represent (or approximate) the properties of the nucleus.

2.2   Intrinsic properties of comet nuclei

The activity of comets at large heliocentric distances is far more revealing of their structure than that close to the sun, which is determined by absorption of solar radiation resulting in sublimation at the very surface of the nucleus. At large heliocentric distances the surface is cold and the activity is driven from within, even if the interior had been heated by a penetrating heat wave originating in insolation. The thermal drag and the effect of this heat wave are strong functions of the internal properties of the nucleus (or, at least, of the deep subsurface layers). What do we know about the activity of distant comets? We may choose as representative examples the puzzling behavior of 2060 Chiron, which stays at a large heliocentric distance throughout its orbit; the sudden outburst detected for comet P/Halley at 14 AU post-perihelion and the very early activity of comet Hale-Bopp, at 7 AU pre-perihelion. Although very different in nature, it appears that all these phenomena may be explained by the same basic model — involving crystallization of porous amorphous ice followed by gas release and flow to the surface — with a different combination of parameters in each case (see Prialnik 1998 and references therein).

We thus conclude that cometary ice must be porous — a characteristic which is independently derived from completely different considerations, e.g., based on estimates of the mass and the volume (which yield an average density that is lower than that of each constituent). The activity pattern of distant comets also provides strong evidence that the ice should be amorphous, trapping considerable amounts of gas, as indicated by experimental studies (e.g., Bar-Nun et al. 1988). Finally, models show that both structure and composition are far from homogeneous; rather, a stratified configuration of the outer layers is obtained, with species of greater volatility more abundant at increasingly larger depths (e.g., Fanale and Salvail 1997). Obviously, the latter properties result from alteration induced by evolution. But the former — the porosity of the nucleus and the amorphous nature of cometary ice — may be taken as typical of the initial structure. May we assume this structure to have prevailed throughout the nucleus, that is, to have survived radiogenic heating?

3 Radiogenic Heating of Comets

3.1 The ^{26}Al source

The radioactive isotope ^{26}Al has long been recognized as a potential heat source capable of melting bodies of radii between 100 and 1000 km (Urey 1955). Its lifetime is just long enough for it to outlive the formation of comets, and short enough to generate sufficient heat power for overcoming cooling by conduction. Renewed interest in this radionuclide followed the detection of interstellar 1.809 Mev γ-rays from the decay of ^{26}Al (Mahoney et al. 1984, Share et al. 1985, Clayton and Leising 1987). Finally, ^{26}Mg enhanced abundances are found in Ca-Al inclusions of meteorites, the most notorious among them being the Allende. All this evidence points towards an interstellar isotopic ratio ^{26}Al/^{27}Al $\approx 5 \times 10^{-5}$, implying an initial mass fraction $X_0(^{26}$Al$) \approx 7 \times 10^{-7}$ in dust and presumably an order of magnitude less in newly formed comets, if their time of aggregation did not exceed a few million years. In conclusion, the radiogenic heat sources available in comets are the usual ^{40}K, ^{232}Th, ^{238}U, ^{235}U, and for a short, but important, period of time ^{26}Al. The corresponding data is listed in Table 1.

3.2 Analytical considerations

In order to assess the potential effect of radiogenic heating during the early evolution of comets, we list in Table 2 the energy per unit mass required by different processes (to be compared with the available energy listed in Table 1). Since the initial ^{26}Al abundance is uncertain, we also list the mass fraction of ^{26}Al that would be required in order for it to supply this energy by itself. We conclude that none of the radioactive isotopes would be capable of evaporating the comet nucleus, but most of them supply sufficient energy for melting the ice or evaporating the other volatile species. However, if the radioactive energy is deposited at a low rate, its effect is bound to be far less significant, since the heat might be lost by radiation at the surface.

A more realistic estimate of the effect of radiogenic heating is obtained from global energy considerations. If we neglect crystallization, as well as internal sublimation and gas flow, and integrate the energy equation (4) over

Table 1. Properties of radioactive isotopes

Isotope	τ	X_0	H	$X_0 H$	$\tau^{-1} X_0 H$
	(years)		(erg g^{-1})	(erg g^{-1})	(erg g^{-1} s^{-1})
^{40}K	1.82(9)	1.1(-6)	1.72(16)	1.89(10)	3.3(-7)
^{232}Th	2.00(10)	5.5(-8)	1.65(17)	9.08(9)	1.4(-8)
^{238}U	6.50(9)	2.2(-8)	1.92(17)	4.22(9)	2.1(-8)
^{235}U	1.03(9)	6.3(-9)	1.86(17)	1.17(9)	3.6(-8)
^{26}Al	1.06(6)	$\sim$5(-8)	1.48(17)	$\sim$7(9)	2.1(-4)

Table 2. Energy required for different processes

Process	T	Δu	H	X_0
	(K)	(erg g^{-1})	(erg g^{-1})	(^{26}Al)
H_2O sublimation	180	1.4(9)	2.8(10)	2.0(-7)
CO_2 sublimation	100	4.5(8)	5.9(9)	4.3(-8)
CO sublimation	30	5.0(7)	2.3(9)	1.6(-8)
H_2O melting	273	3.0(9)	3.3(9)	4.3(-8)
H_2O crystallization	130	7.4(8)	- 9(8)	5.0(-9)

the entire nucleus, we have:

$$\int \frac{\partial u}{\partial t} dm = M \sum_j \tau_j^{-1} X_{0,j} H_j e^{-t/\tau_j} - 4\pi R^2 \left[\sigma T(R)^4 - \frac{(1-A)L_\odot}{16\pi d_H^2} \right]. \quad (7)$$

Defining an equilibrium surface temperature

$$T_{eq}(d_H) = \left[\frac{(1-A)L_\odot}{16\pi d_H^2 \sigma} \right]^{1/4}, \quad (8)$$

we obtain

$$\langle \frac{du}{dt} \rangle = \sum_j \tau_j^{-1} X_{0,j} H_j e^{-t/\tau_j} - \frac{3\sigma}{R\rho}(T_s^4 - T_{eq}^4). \quad (9)$$

Now, the difference between T_{eq} and the actual surface temperature $T(R)$ is due to heat conduction into or out of the nucleus. Roughly,

$$\sigma T_s^4 - \sigma T_{eq}^4 = -K\frac{dT}{dr} \sim K\frac{T}{R}. \quad (10)$$

Hence

$$\langle\frac{du}{dt}\rangle = \sum_j \tau_j^{-1} X_{0,j} H_j e^{-t/\tau_j} - \frac{3KT}{R^2\rho}. \tag{11}$$

The maximal heating rate, obtained at $t = 0$, is $\tau_j^{-1} X_{0,j} H_j$; the numerical values of this factor for each radiogenic species are listed in Table 1. The cooling rate,

$$\frac{3KT}{R^2\rho} \approx \frac{0.0016}{R_{km}^2} \text{ erg g}^{-1}\text{ s}^{-1} \tag{12}$$

for amorphous ice , is lower than the rate of heating for $R \gtrsim 3$ km in the presence of ^{26}Al, and for $R \gtrsim 70$ km in its absence. For crystalline ice the cooling rate is about 20 times higher and hence higher cometary radii are required for internal heating to take place. As the heating rate declines with time, it eventually becomes lower than the rate of cooling in all cases. In conclusion, for suitably large comets the internal temperature rises at the beginning up to a maximal value T_{max} and then falls, tending to T_{eq}. Obviously, a larger radius or a higher ^{26}Al content would lead to a higher T_{max}. The factor K/ρ in eq.(12) decreases with increasing porosity ($\psi(p) < 1 - p$) and hence T_{max} varies as the porosity. However, the presence of volatiles in the interior may reverse this trend, since gas flow, which increases with porosity, may contribute significantly to heat transfer. In addition, the pressure that gases exert could, in turn, affect the porous structure.

3.3   Time Scales

The evolution of a comet is thus determined by the competition between the rates of several different processes, or by the relationship between their corresponding time scales, as plotted in Figure 1: (a) the thermal time scale, where distinction must be made between the different components, particularly between amorphous ice, which is a poor heat conductor, and crystalline ice; (b) the time scale of gas diffusion, and hence – pressure release; (c) the time scale of crystallization, and hence – gas-release and pressure build-up; (d) the time scales of sublimation of the different volatiles; (e) the time scales of radioactive decay. We now consider the possible effect of radioactive heating, mainly by ^{26}Al (see Prialnik & Podolak 1995). At a depth of 1 km the decay time of ^{26}Al becomes comparable to the thermal time scale of amorphous ice, meaning that the ice might be barely heated; it would certainly be heated at larger depths, a few km and beyond. Eventually, the internal temperature would become sufficiently high for crystallization to set in, providing an additional internal heat source. At the same time, however, the thermal time scale would decrease, crystalline ice being a much better heat conductor than amorphous ice. Hence, only in still larger comet nuclei (beyond 10 km) would the internal temperature continue to rise. If the internal temperature becomes such that the time scale of sublimation is shorter than the time scale

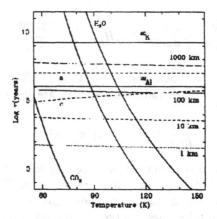

Fig. 1. Time scales *vs.* temperature: gas diffusion time scales are given for different depths (comet radii), as marked; the dashed curves are the thermal time scales for amorphous ice (a) and crystalline ice (c), for a depth of 10 km; the unmarked line corresponds to crystallization

of radiogenic heat release, then most of the released energy will be absorbed in sublimation of ice from the pore walls. If, in addition, the radius is such that the time scale of gas (vapor) diffusion is lower than the time scale of sublimation, then sublimation will consume the radiogenic heat indefinitely (as long as there is ice), since the vapor will be efficiently removed. A steady state will develop, without further heating of the ice matrix. It is worth mentioning that the temperature of such a steady state would be considerably lower than the melting temperature of ice. On the other hand, if the porosity of the ice is very low and the average pore size is very small, the gas diffusion time may become sufficiently high for gas removal to become inefficient. In such cases the internal temperature might rise to the melting point of ice.

Analytical considerations alone cannot lead to any definite conclusion regarding the effect of radiogenic heating of comets, although they show that it could be significant. We thus resort to the results of numerical computations.

4   Early Evolution of Comets Due to Radiogenic Heating

4.1   Non-porous comet nuclei

We first consider the behavior of a non–porous icy sphere. By 'non–porous' we mean that gases cannot be produced or flow in the interior of the nucleus; but we allow for the reduction in thermal conductivity due to porosity. A 20 km radius comet nucleus made of a non–porous mixture of amorphous ice and dust (in equal mass fractions) was studied by Prialnik and Podolak

(1995), assuming an initial ^{26}Al mass fraction of 5×10^{-8}. As the central temperature exceeds $\sim$ 70 K, crystallization starts at a very slow rate. The latent heat released accelerates the process, which reaches its peak at $\sim$ 130 K. As the crystallization front advances through the nucleus, the crystalline core continues to heat up, until the melting temperature is reached. A liquid core, extending out to a radius of 12 km, is maintained for several million years. At the end of the crystallization process, only a negligibly thin outer layer is left in its amorphous form. When the ^{26}Al is exhausted, and the temperatures throughout the nucleus decline, refreezing starts at the outer boundary of the liquid core and recedes toward the center. Finally, the nucleus cools off to a very low and almost uniform temperature (due to the high conductivity of crystalline ice). Larger bodies require lesser amounts of ^{26}Al for achieving similar results. Beyond $\sim$ 500 km, the other radiogenic species supply sufficient heating power for interior melting (Consolmagno and Lewis 1978, Prialnik and Bar-Nun 1990). If a low conductivity is assumed, comets having radii of only $\sim$ 200 km may reach the melting point of ice without ^{26}Al, as shown by Yabushita (1993).

The conductivity of the ice determines not only whether or not melting is attained at the center, but also the extent of the liquid core and the cooling time scale. In order to test how extended the liquid core could become, Prialnik and Podolak repeated their calculations of the 20 km nucleus, adopting thermal conductivity coefficients reduced by a factor of 1000 (as indicated by Kouchi et al. 1992, although only for amorphous ice). The evolution through the phases of heating, melting, and further heating of the liquid core is found similar to the previous case, but the cooling phases — cooling of the core to the freezing point, freezing, and further cooling of the ice — are considerably longer. The liquid core extends up to a few hundred meters below the surface and is maintained for almost 10^9 years. An outer layer more than 300 m thick is left in the amorphous form. This layer constitutes, in fact, the only difference between the two cases that might be detected at present time, i.e. after $\sim 4.5 \times 10^9$ years of evolution in the remote outskirts of the solar system.

In conclusion, comet nuclei made of a *solid* (non–porous) mixture of ice and dust can reach the melting temperature in their interior and form liquid cores, provided that some ^{26}Al is present in the initial composition. The core may remain liquid for periods of time ranging from 10^6 to 10^9 years, depending on the thermal conductivity of the mixture.

The effect of radiogenic heating of comets combined with a low thermal conductivity was studied in more detail by Haruyama et al. (1993). The conclusion was that even small comets (a few km in radius) may undergo runaway crystallization during residence in the Oort cloud (or the Kuiper belt). Moreover, the internal temperatures of all comets may rise sufficiently, even in the absence of ^{26}Al, for volatile species to be lost from the nucleus or to concentrate towards the surface. However, as we shall show shortly, the flow of gas through the nucleus has a significant effect on the thermal conductivity, which may completely change this conclusion.

4.2 Highly porous comet nuclei

We now consider highly porous material, where gases can flow through the nucleus and sublimation from (or condensation onto) the pore walls may occur. Starting with the relatively simple case of crystalline ice, Prialnik and Podolak (loc. cit.) find that when the ice is porous, the evolutionary course is entirely different from that obtained for a similar configuration, but with solid ice. Even for the same ^{26}Al content, the temperatures obtained are significantly lower, far below the melting point. When the time scale of sublimation, which is strongly temperature dependent, becomes comparable to the time scale of vapor diffusion through the nucleus, a steady state is achieved. Due to the effect of internal sublimation of ice from the pore walls, a flat temperature profile is obtained. A lower conductivity of the ice matrix does not change the peak temperature; we recall that this temperature is determined by the equilibrium between sublimation and vapor flow, both unaffected by the conductivity. Only the cooling time, after the ^{26}Al is exhausted, is much longer, as expected, since this time is mainly determined by the thermal conductivity of the solid matrix.

If the initial ^{26}Al mass fraction is increased even by a factor of 10, the peak temperature reached is barely affected (for the same reason as before), but the rise in temperature occurs sooner, due to the increased rate of radioactive energy release. A different result may be expected if the porosity is decreased considerably still adopting the high ^{26}Al abundance. Since the vapor diffusion time scale is inversely proportional to the porosity, steady state should be reached in this case at a higher temperature. The same effect should be obtained by increasing the cometary radius, which would also increase the gas diffusion time scale. These cases are considered in the next section.

We conclude that if comet nuclei are made of *porous crystalline ice*, they do not reach high temperatures, in spite of radioactive heat release (even in the presence of ^{26}Al), and they do not develop liquid cores at any time in their early history.

We now turn to amorphous porous ice and consider first models that do not include ^{26}Al. For radii below 20 km, the central temperature rises several tens of degrees during a period of $\sim 10^8$ years, and then declines gradually. The amount of ice transformed from the amorphous to the crystalline phase in these bodies is negligibly small, as is the mass of occluded gas released. For radii $R \geq 25$ km, however, the nucleus is sufficiently large for the thermal diffusion time scale from the center to the surface to exceed the heating time scale of the radioactive elements. As a result, the central temperature rises to levels where the transformation from amorphous to crystalline ice can proceed rapidly. The latent heat released drives the crystallization front further towards the surface.

However, since the body is porous, an additional cooling mechanism is available. The process of crystallization releases trapped gas, and this outflowing gas carries the heat away rapidly. The competition between heating by latent heat release and cooling by the flowing gas determines the depth at which the crystallization front stops. A fraction of the outflowing gas condenses in the cooler outer layers of the nucleus, just outside the region of crystalline ice, forming a shell with an enhanced content of ice different from H_2O. The structure of such bodies can therefore be described as consisting of three layers: a core of crystalline ice, a shell of amorphous ice enriched in ices of other volatiles, and an outer mantle of essentially pristine composition. Such a layered composition resulting from internal heating (shown schematically in Figure 2) had long been suggested by Whipple and Stefanik (1966) and speculated on years later by Yabushita and Wada (1988) and by Yabushita (1993). A very thin upper crust is significantly altered by exterior sources, such as UV radiation and cosmic rays (cf. Stern 1998). The range

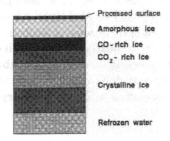

Fig. 2. Schematic representation of a layered structure resulting from early evolution (arbitrary scale): note that the crystalline ice is arbitrarily divided, to indicate different porous structures (the porous structure changes smoothly)

of cometary radii where such a configuration is encountered is, however, restricted. It requires a sufficiently hot interior to produce gas at a significant rate and, at the same time, a sufficiently deep cold outer layer for the gas to freeze. If the cold outer layer (below the surface) is too thin the gas pressure there will be too low to produce condensation. A large, porous comet nucleus crystallizes almost completely during a very short period of time. Thereafter the temperature drops sharply over the entire nucleus, due to the relatively high conductivity of crystalline ice. Only an outer layer of a few hundred meters thickness remains amorphous and no ice besides H_2O survives in this case.

The situation changes when ^{26}Al is introduced. Crystallization sets in much sooner, on time scales of the order of the lifetime of ^{26}Al. For a small amount of ^{26}Al, crystallization is still negligible for $R \leq 15$ km, where cooling through the surface is efficient enough to keep the internal temperature below the crystallization threshold. For $R > 15$ km, cooling is less efficient, and crystallization takes place throughout a central region that extends out to about half or more the radius (1/8 or more of the mass). The temperature profile (and hence the crystalline ice mass fraction) is maintained flat due to heat advection by the flowing gas, which surpasses conduction by the solid matrix. As before, for a restricted range of radii a layer enriched in ice of the trapped volatile species forms at some depth below the surface. Except for the shift of the critical radii to lower values, the only difference between evolutionary trends with or without ^{26}Al is that the peak temperature reached in the latter case is somewhat higher and occurs much sooner. After $\sim 10^8$ years both comet models are equally cold. Moreover, both bear the marks of high internal gas pressures that occurred at the time of crystallization. The initial structure of a homogeneous porous material with very small pores (grains), has changed. The pores have been enlarged and they are no longer uniform: pore sizes increase from the center outwards and decrease again towards the surface. However, the average pore size is larger for the nucleus that has undergone crystallization (and gas release) at a more intense rate, due to the presence of ^{26}Al. The amorphous ice outer layer is much thinner in this case.

In conclusion, bodies made of *porous amorphous ice and dust* undergo crystallization due to radioactive heating, if their radii exceed a critical value. The critical radius is determined by the initial ^{26}Al abundance. The progress of the crystallization process is determined by the amount of occluded gas (e.g., CO), while the formation of a CO ice shell depends on the radius of the icy body.

4.3   Large comet nuclei of low porosity

Prialnik and Podolak (1998) have considered a large comet, 100 km in radius, composed of water ice and dust in equal mass fractions, moving in a circular orbit at a distance of 100 AU from the sun (approximately representing a Kuiper belt object). The porosity was taken to be 0.1 and the average pore size, 100μm. Small fractions of CO and CO_2 gases were assumed to be initially occluded in the amorphous ice. The evolution of this model was followed through the rise to T_{max} and decline from it, adopting an initial ^{26}Al mass fraction of 5×10^{-8}.

At the beginning, as the radiogenic heat release is far more efficient than conduction, the temperature rises steadily. With it, the rate of crystallization rises, and since this is an exothermic process, it escalates into a runaway. Eventually, the temperature becomes sufficiently high for sublimation to set in and the rise in temperature comes to a stop when the rate of energy release roughly equals the rate of absorption due to sublimation (part of the energy is transferred to the surface). As the rate of sublimation is hindered by the build-up of internal vapor pressure, the peak temperature attained is quite high, ~ 260 K. The high (total) internal pressure breaks the pores and thus the average pore size grows. As the tensile strength decreases with radial distance, the originally uniform pore size of 100μm increases gradually to a few mm, starting at a depth of ~ 70 km outwards. Due to the excellent (effective) thermal conductivity of vapor filled porous ice (see e.g., Prialnik 1992) a flat temperature profile is obtained throughout a large inner part of the nucleus. Farther out, the temperature declines; the surface temperature is higher than T_{eq}. The CO_2 gas released in the interior freezes when it encounters ia temperature of about 100 K. Another temperature plateau forms at a temperature slightly above 100 K, maintained by the CO_2 thermostat. This process is transient, however, and gradually the CO_2 ice evaporates.

At the adopted heliocentric distance of 100 AU, the temperature is not sufficiently low for CO to freeze as well. In the end, both the CO and the CO_2 are completely lost. Thereafter, the temperature profile is uniquely determined by the H_2O and the ~ 260 K plateau extends up to about 15 km from the surface. At this depth evaporation reaches a peak, since the temperature is high, but the pressure drops sharply towards the surface and the vapor is efficiently removed. Thus, although in the deep interior the rate of sublimation is too low to alter the porosity significantly, at a depth of ~ 15 km, due to vigorous ice sublimation, a layer of relatively high prosity and low water ice content is formed. Above this layer, which is a few kilometers thick, the temperature is too low for sublimation to be of importance and the initial porosity and ice mass fraction are maintained. The formation of such a 'weak' zone at some depth below the surface may have implications for the future history of the comet. Although the temperature and pressure will decline when the radiogenic energy source will be exhausted, the altered porous structure will be preserved.

The evolutionary course described above is not necessarily typical of comets in general. In order to obtain a broader picture, Prialnik and Podolak have analyzed the effect of the initial parameters. The maximal temperature attained as a function of porosity is shown in Figure 3 for different cometary radii. Generally, the temperature rises at a rate determined by the competition between heat release and conduction, until steady state is achieved between heat release by radioactive decay on the one hand and conduction and heat absorption by sublimation on the other hand. For the small nucleus, the peak temperature attained increases with the porosity, reflecting the effect of the porosity on the thermal conductivity of the solid matrix. For the large nucleus an opposite effect is encountered: the peak temperature decreases as the porosity increases. This is due to the fact that at the high temperatures attained the ice crystallizes and releases the trapped gas. Heat transfer is now due predominantly to advection by the flowing gases and the advection rate obviously increases with porosity. For the intermediate size of 10 km the behavior changes trend at a porosity of about half. The fractional mass throughout which the temperature is close to T_{max}, as well as the time required to reach a steady configuration, depend on the initial ^{26}Al abundance. A *threshold* initial abundance exists, of $\sim 10^{-8}$, below which the effect of ^{26}Al is almost negligible, and above which it is considerable, but does not increase appreciably with a further increase of $X_0(^{26}$Al$)$. The same effect was noted by Prialnik and Bar-Nun (1990). As a function of pore size, the maximal temperature peaks at $10\mu m$ and decreases for larger pores due to the increase in gas flow (and advection). If the average pore size is smaller than $10\mu m$, however, the temperature decreases again. This is due to the large surface to volume ratio that is obtained for small pores, which enhances the rate of sublimation. Finally, the amount of occluded gas affects the effective conductivity considerably: between $f \approx 0.02$ and $f \approx 0.1$ the peak temperature drops from over 260 K to less than 100 K. In the cases when the temperature remains low, so does the pressure, and pore breaking is less severe. The outflowing gases refreeze in the outer cold layers of the nucleus, closing the pores to some extent and leading to a layered composition.

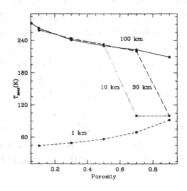

Fig. 3. Maximal temperature as a function of porosity, for different nucleus radii

5   Conclusion: Characteristics of Distant (New) Comets

Porous cometary nuclei composed initially of amorphous ice and dust retain the ice in the amorphous form if their radii do not exceed 20 km, and if the initial ^{26}Al content is negligible ($\leq 5 \times 10^{-9}$). Obviously, no liquid water is ever produced under these circumstances. If the radius is significantly larger than 20 km, the nucleus crystallizes even in the absence of ^{26}Al and only a thin outer layer preserves the pristine composition. The thickness of this layer depends on the thermal conductivity of the ice. The presence of ^{26}Al in the initial composition reduces the critical radius for crystallization. [Porous crystalline nuclei do not reach the melting temperature, even under extremely favorable conditions, such as very low conductivity and very abundant ^{26}Al.]

Thus porous comet nuclei may emerge from the long–term evolution in the outskirts of the solar system in three different configurations: a) preserving their pristine structure throughout; b) almost completely crystallized (except for a relatively thin outer layer) and, possibly, with an altered pore structure, the average pore size increasing towards the surface, and c) having a crystallized core, layers of frozen gases (originally occluded in the amorphous ice), the more volatile species closer to the surface, and an outer layer of unaltered pristine material. The outcome and the details of the final configuration are determined by initial conditions, such as radius and composition, as well as by the properties of cometary material. The layered structure is the same — only in reverse and on a larger scale — as that obtained below the surface when the comet is heated by the sun and the heat wave propagates from the surface inwards, rather than from the center outwards (see Section 2.1).

Liquid cores may be obtained if the porosity is very low and the comet very large. The larger the comet, the higher is the upper limit on the porosity for which water may be obtained. The extent of liquid cores and the length of time during which they remain liquid are again determined by initial conditions, as well as by physical properties of the ice. Obviously, the refrozen ice is crystalline. Only for a negligible porosity and a very low conductivity (requirements which are apparently incompatible), is it possible to have both an extended liquid core, for a considerable period of time, and an outer layer of significant thickness that has retained its original pristine structure.

Thus, we may answer the opening questions as follows: very small comets (which are not fragments of larger ones) are indeed pristine objects; very large comets have probably had liquid cores in the past, in which prebiotic molecules (amino acids) may have formed, to be dispersed later by impacts throughout the solar system; some comets, those in the intermediate size range, may have melted in the center, while retaining the outer layers in original, unaltered, form. Unfortunately, given the uncertainties related to the properties of cometary ice and to the formation of comets, we cannot be more specific than that.

References

1. Bar-Nun, A., Kochavi, E., and D. Laufer (1988) Trapping of gaseous mixtures by amorphous water ice. Phys. Rev. B. **38**, 7749–7754

2. Benkhoff, J., and W. F. Huebner (1995) Influence of the vapor flux on temperature, density and abundance distributions in a multi-component, porous, icy body. Icarus **114**, 348–354

3. Clayton, D. D. and M. D. Leising (1987) ^{26}Al in the interstellar medium. Phys. Rept. **144(1)**, 1–50

4. Consolmagno, G. J., and J. S. Lewis (1978) The evolution of icy satellite interiors and surfaces. Icarus **34**, 280–293

5. Espinasse, S., A. Coradini, M. T. Capria, F. Capaccioni, R. Orosei, M. Salomone, and C. Federico (1993) Planet. Space Sci. **41**, 409–

6. Fanale, F., and J. R. Salvail (1997) The cometary activity of Chiron: a stratigraphic model. Icarus **125**, 397–405

7. Greenberg, J. M. (1998) Making a comet nucleus. Astron. Astrophys. **330**, 375–380

8. Haruyama, J., Yamamoto, T., Mitzutani H., and J.M. Greenberg (1993) Thermal history of comets during residence in the Oort cloud: effect of radiogenic heating in combination with the very low thermal conductivity of amorphous ice. J. Geophys. Res. **98**, 15079–15088

9. Irvine, W. M., Leschine, S. B., and F. P. Schloerb (1980) Thermal history, chemical composition and relationship of comets to the origin of life. Nature **283**, 748–749

10. Kouchi, A., Greenberg, J. M., Yamamoto, T., and T. Mukai (1992) Extremely low thermal conductivity of amorphous ice: relevance to comet evolution. Astrophys. J. **388**, L73–L76

11. Mahoney, W. A., Ling, J. C., Wheaton, Wm. A., and A. S. Jacobson (1984) *Heao 3* discovery of ^{26}Al in the interstellar medium. Astrophys. J. **286**, 578–585

12. Mekler, Y., and M. Podolak (1994) Formation of amorphous ice in the protoplanetary nebula. Planet. Space Sci. **42**, 865–870

13. Mekler, Y., D. Prialnik, and M. Podolak (1990) Evaporation from a porous comet nucleus. Astrophys. J. **356**, 682–686

14. Podolak, M., and D. Prialnik (1997) ^{26}Al and liquid water environments in comet. in Comets and the Origin of Life (P. Thomas, C. Chyba, and C. McKay, eds.) Springer-Verlag, New York, pp. 259–272

15. Prialnik, D. (1992) Crystallization, sublimation, and gas release in the interior of a porous comet nucleus. Astrophys. J. **388**, 196–202

16. Prialnik, D. (1998) Modelling gas and dust release from Comet Hale-Bopp. Proceedings of The First International Conference on Comet Hale-Bopp (in press)

17. Prialnik, D., and A. Bar-Nun (1990) Heating and melting of small icy satellites by the decay of ^{26}Al. Astrophys. J. **355**, 281–286

18. Prialnik, D., A. Bar–Nun, and M. Podolak (1987) Radiogenic heating of comets by ^{26}Al and implications for their time of formation. Astrophys. J. **319**, 992–1002

19. Prialnik, D., Egozi, U., Bar–Nun, A., Podolak, M., and Y. Greenzweig (1993) On pore size and fracture in gas–laden comet nuclei. Icarus **106**, 499–507

20. Prialnik, D., and M. Podolak (1995) Radioactive heating of porous comet nuclei. Icarus **117**, 420–430

21. Schmitt, B., S. Espinasse, R. J. A. Grin, J. M. Greenberg, and J. Klinger (1989) Laboratory studies of cometary ice analogues. ESA–SP **302**, 65–69

22. Share, G. H., Kinzer, R. L., Kurfess, J. D., Forrest, D. J., Chupp, E. L., and E. Rieger (1985) Detection of galactic ^{26}Al gamma radiation by the *SMM* spectrometer. Astrophys. J. **292**, L61–L65

23. Steiner, G., and N. I. Kömle (1991) Thermal budget of multicomponent porous ices. J. Geophys. Res. **96**, 18897–18902

24. Stern, S. A. (1998) Evolutionary processes affecting comets in the Oort cloud and the Kuiper belt. IAU Proceedings (in press)

25. Tancredi, G., Rickman, H., and J. M. Greenberg (1994) Thermochemistry of comet nuclei. Astron. Astrophys. **286**, 659–682

26. Urey, H. C. (1955) Proc. Nat. Acad. Sci. **41**, 127–

27. Wallis, M. K. (1980) Radiogenic heating of primordial comet interiors. Nature **284**, 431–433

28. Whipple, F. L., and R. P. Stefanik (1966) On the physics and splitting of cometary nuclei. Mem. Roy. Soc. Liege (Ser. 5). **12**, 33–52

29. Yabushita, S. (1993) Thermal evolution of cometary nuclei by radioactive heating and possible formation of organic chemicals. Mon. Not. R. Astron. Soc. **260**, 819–825

30. Yabushita, S., and K. Wada (1988) Earth, Moon and Planets **40**, 303–

Optical Observations of Trans-Neptunian and Centaur Objects

M. Antonietta Barucci[1] and Monica Lazzarin[2]

[1] Observatoire de Paris, 92195 Meudon Principal Cedex, France,
   e-mail: barucci@obspm.fr
[2] Dipartimento di Astronomia, 35122 Padova, Italy, e-mail: lazzarin@pd.astro.it

Abstract. Trans-Neptunian objects are probably primitive remnants from the early accretional phases of the solar system and may contain the most pristine unprocessed material. To investigate the surface properties of these objects we started an observational campaign at the European Southern Observatory (La Silla, Chile) with the 3.6 m telescope and the 3.5 m New Technological Telescope (NTT). We observed U, B, V, R, I colors for six Trans-Neptunian objects (1994 JR$_1$, 1994 TB, 1995 QY$_9$, 1996 TL$_{66}$, 1996 TO$_{66}$, and 1996 TP$_{66}$) and visible spectra for five Centaurs (2060 Chiron, 7066 Nessus, 8405 1995 GO, 1995 DW$_2$, 1997 CU$_{26}$). The basic result is that the two classes of objects are remarkably similar, even if a great diversity exists among their colors implying a wide range of surface compositions.

1 Introduction

Speculation on the existence of a trans-Neptunian disk of icy objects dates back to the 1940s. In 1943, Edgeworth suggested the existence of a large number of small bodies beyond the orbits of the planets and later on, Kuiper (1951) suggested an alternative source of short period comets just beyond the planets. Only in 1992, with the discovery of 1992 QB$_1$ (Jewitt and Luu, 1993), have astronomers become aware that a vast population of small bodies orbits the Sun beyond Neptune. To date about 70 Trans-Neptunian objects have been discovered (Marsden, 1998), but probably more than 100,000 bodies (Jewitt, 1998) exist with diameter larger than 100 km in a radial zone extending outwards from the orbit of Neptune (30 AU) and 50 AU.

The Centaurs are a dynamically separate family of objects on unstable orbits whose semimajor axes fall between those of Jupiter and Neptune, with a dynamical lifetime measured in millions of years. Long term orbital integrations of the Trans-Neptunian objects suggest that perturbations by Uranus and Neptune provide a source of Centaurs and short period comets (Levison and Duncan, 1997 and Morbidelli, 1997). Eight Centaurs have been identified up to now even if Jedicke and Herron (1997) affirmed that in the absolute magnitude H range -4 to 10.5 there must be fewer than about 2000 Centaurs implying a population large as the Main Belt asteroids. (320 KBOs and 20 Centaurs are known as of July 2000. The number of Centaurs larger than 100 km diameter is inferred by Sheppard et al. (2000) Astronomical Journal, in press, to be near 100 - Eds).

All these objects are supposed to be the remnants from the accretion disk which formed the solar system. They might be composed by the least thermally processed materials in the solar system, while their evolution has been influenced by resonances and collisions.

The study of these bodies has rapidly evolved in the last few years with discoveries of new objects and theoretical advances. Because of their small sizes and large heliocentric distances, they are faint and difficult to study: therefore very little is known on their physical and compositional properties (Davies, 1998). Near-Infrared spectra are available only for two Trans-Neptunian objects and three Centaurs, each of them showing different behaviour which implies different surfaces. Tegler and Romanishin (1998) analysing the BVR colors of 13 objects (9 Trans-Neptunian and 4 Centaurs) found two distinct groups: one neutral and the other very red. To investigate the compositional properties of these objects, we started an observational campaign of Trans-Neptunian objects and Centaurs, since the latter are believed to have escaped from the Kuiper Belt.

2   Observations

We started the observational campaign in 1997 at La Silla observatory (ESO, Chile). We observed five Centaurs (Table 1) with the 3.6 m telescope using EFOSC (ESO Faint Objects Spectrograph and Camera) as spectrograph. Six Trans-Neptunian objects (Table 2) have been observed with the 3.5 m New Technology Telescope (NTT) using the direct imaging CCD camera SUSI2 and the Bessel filters: U, B, V, R, and I. Due to the faintness of these last ones, and the fact that their orbits are not well determined, it was easier to observe them in imaging and to obtain broad band spectra.

The observations of these faint objects require a great attention. The data reduction has been performed using the software package MIDAS and IRAF and we pay particular attention on the background contamination and also in taking into account all the sources of errors. For the calibration we observed standard stars from Landolt (1992) and solar analogs from Hardorp (1978).

3   Results

We obtained visible spectra for five Centaurs (2060 Chiron, 7066 Nessus, 8405 1995 GO, 1995 DW_2 and 1997 CU_{26}) and we obtained broad band spectra for six Trans-Neptunian objects (1995 QY_9, 1996 TL_{66}, 1996 TO_{66}, 1994 TB, 1994 JR_1, and 1996 TP_{66}) (Fig.1). The reflectivity of the Trans-Neptunian objects has been computed using the solar colors (Hardorp, 1980) while for the spectra of the Centaurs, several solar analog spectra (Hardorp, 1978), observed in the same nights, have been used. We find a wide dispersion in the optical color within each class of objects and a great similarity between the two classes. The spectra of the Centaurs have been analysed (Lazzarin and

Table 1. Aspect data of the observed Centaur objects: object, date of observations, heliocentric distance, geocentric distance and phase angle

Observed Centaurs	Date [UT]	r [AU]	Δ [AU]	α deg
2060 Chiron	1998/03/29	8.90	8.13	4.22
7066 Nessus	1998/03/30	14.17	13.88	3.90
8405 1995 GO	1998/03/30	10.23	9.34	2.58
1995 DW2	1998/03/29	18.97	17.99	0.57
1997 CU26	1997/03/29	13.67	13.10	3.50

Table 2. Aspect data of the observed Trans-Neptunian objects: object, date of observations, heliocentric distance, geocentric distance and phase angle

Observed Kuiper objects	Date [UT]	r [AU]	Δ [AU]	α deg
1994 JR$_1$	1997/08/28	34.76	34.68	1.66
1994 TB	1997/08/28	30.42	29.50	0.80
1995 QY$_9$	1997/08/28	29.68	28.68	0.36
1996 TL$_{66}$	1997/08/26	35.16	34.69	1.46
	1997/0828	35.16	34.67	1.45
1996 TO$_{66}$	1997/08/27	45.75	44.84	0.56
	1997/08/28	45.75	44.84	0.54
1996TP$_{66}$	1997/08/26	26.44	25.81	1.74
	1997/08/28	26.44	25.78	1.69

Barucci, 1998 and Barucci et al. 1998b) to look for weak cometary emission features, in particular the CN band emission at 3880 Å, but no CN emission feature has been detected within 3-sigma on any of the investigated spectra.

The differences in the spectral behaviour of both population imply a diversity in the surface composition of these objects as it was already found by Luu and Jewitt (1996) and Green et al. (1997). The difference in color is probably due to surfaces with different cosmic radiation exposure due to different degree of collisional alteration revealing fresh interior material. No correlation has been found between colors and orbital distances.

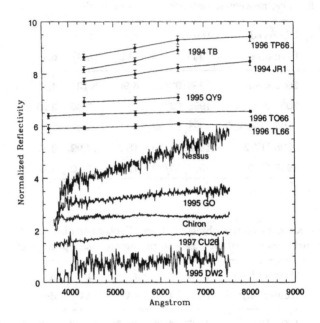

Fig. 1. Normalized reflectivity vs. wavelengh for the six observed Trans-Neptunian objects and the five observed Centaurs. The spectra are vertically offset for clarity, but preserve a fixed scale.

4   Conclusions

The obtained spectra show a wide range of types from flat to very red which imply a wide range of surface compositions. The obtained colors are very similar between the Centaurs and the Kuiper objects. This compositional similarity supports the hypothesis that Centaurs are Trans-Neptunian objects injected into giant-planet crossing orbits and represent transition objects between the Kuiper Belt objects and comet nuclei.

It is not possible to make unique compositional diagnostics based on broadband colors alone or on optical spectra. The neutral colors are compatible with a wide range of surface materials including dirty water ice, while the reddest can have carbon rich compound or organics on the surface. The fact that a difference in the spectra implies a diversity in the surface composition was also pointed out by Barucci et al. (1998a) who analyzed all published colours up to now. Their analysis on the B, V, R, and I colors of 22 Trans-Neptunian objects did not confirm the bimodality found by Tegler and Romanishin (1998) implying that the population of Kuiper objects seems much more complex. In fact each new added datum modifies significantly the previous scenario, probably indicating a finer structure of a complex and inhomogeneous population. Erosion could dominate over accretion, but the question on how these objects formed is still open.

Presently, the level of understanding of these distant objects is similar to that concerning asteroids at the beginning of the 60's, when the few known asteroids were divided into two groups on the basis of their UBV colors. The qualitative and quantitative increase of data allowed researchers to draw a quite complete scenario for the asteroids. Thus there is a strong need of high quality observational data on this new population. The faintness of these objects represents a formidable challenge for the astronomers and only the use of 8/10 m class telescopes will allow to obtain a more precise picture of these distant populations.

References

1. Barucci, M. A., Doressoundiram, A., Tholen D., Fulchignoni M., Lazzarin M. (1998a) Spectrophotometric observations of Edgeworth-Kuiper Belt objects. Icarus, submitted

2. Barucci, M. A., Lazzarin M, Tozzi G. P (1998b) Compositional surface variety among the Centaurs. Astron. Jour., submitted

3. Davies, J. K. (2000) Physical Characteristics of Trans-Neptunian Objects and Centaurs. These proceedings

4. Edgeworth, K. E. (1943) The evolution of our planetary system. J.Br. Astron. Assoc. **53**, 181–188

5. Green, S. F., McBride, N. et al. (1997) Surface reflectance properties of distant Solar system bodies. MNRAS **290**, 186–192

6. Hardorp, J., (1978) The sun among stars. Astron. Astrophys. **63**, 383–390

7. Hardorp, J., (1980) The sun among stars. Astron. Astrophys. **91**, 221–232

8. Jedicke, R. , J. D. Herron (1997) Observational Constraints on the Centaur Population. Icarus **127**, 494–507

9. Jewitt, D. C., (2000) The Kuiper Belt: Overview. These proceedings.

10. Jewitt, D. C., Luu J. X. (1993) Discovery of the candidate Kuiper belt object 1992 QB$_1$, Nature **362**, 730–732

11. Kuiper, G. P. (1951) On the origin of the Solar System. In Astrophysics: A total Sympositum (J.A. Hynek, Ed.) pp.357–424. McGraw-Hill, New York

12. Landolt, A. (1992) UBVRI Photometric standard stars in the magnitude range $11.5 < V < 16.0$ around the celestial equator. Astron. J. **104**, 340–371

13. Lazzarin, M., and Barucci, M. A., (1998) Spectroscopic investigation of the Centaurs. B.A.A.S. **30**, 1114

14. Levison, H. F., and Duncan, M. J., (1997) From the Kuiper Belt to Jupiter-family Comets: the spatial distribution of ecliptic comets. Icarus **127**, 13–32

15. Luu, J., Jewitt, D., (1996) Color diversity among the Centaurs and Kuiper belt objects Astron. J. **112**, 2310–2318

16. Marsden, B. G. (1998) List of Transneptunian Objects (1998) available at http://cfa-www.harvard.edu/cfa/ps/list/TNOs.html

17. Morbidelli A. (1997) Chaotic diffusion and the origin of Comets from 2/3 resonance in the Kuiper belt. Icarus 127, 1–12
18. Tegler S. C., Romanishin W. (1998) Two distinct populations of Kuiper-belt objects. Nature 392, 49–50

Photometry Techniques — Report of Splinter Meeting

Simon Green and Neil McBride

Planetary and Space Science Research Institute, The Open University, Milton Keynes MK7 6AA, UK.
email: s.f.green@open.ac.uk, n.m.mcbride@open.ac.uk.

Abstract. During the MBOSS workshop, a splinter meeting was held to discuss the potential sources of the discrepancies in observer's reported photometry of Kuiper Belt Objects. The meeting focused on data acquisition and reduction techniques. It was agreed that a test of various observers' methods would be undertaken, and the results would be collated and reported at the *Asteroids Comets Meteors 1999* conference.

1 Introduction

It has been clear for some time that the reported results of broadband photometric studies of Kuiper Belt Objects by different authors have been discrepant, with colour differences exceeding the quoted error bounds by a considerable margin. This was highlighted during the sessions on physical properties where new results were presented and compared by the observers present. Figure 1 illustrates the published and unpublished results for two 'well observed' objects and highlights the problem. How can such differences arise and can the photometry be trusted by those trying to interpret the results? It is possible that there are real magnitude changes for some objects (due to activity or rotational lightcurves) but colour variations are likely to be second order effects. In addition most observers adopt an observing strategy to avoid or average out such changes. A splinter session was hastily arranged in suitably relaxed surroundings (the Hotel Am-Park bar), to discuss each participant's observation and data reduction strategy.

KBO's move relatively slowly against the sky background (typically 3 arcmins per hour) so frames are usually taken with sidereal tracking, and exposures are chosen to match the pixel-crossing time ($\sim$100–600 seconds). Repeated frames are taken, usually in colour sequences such as RBRVRIR to provide improved signal to noise, and identify or average out variations due to possible rotational lightcurves. Sequences may be repeated after several hours or on separate nights to identify any effects of changing faint background sources in the aperture.

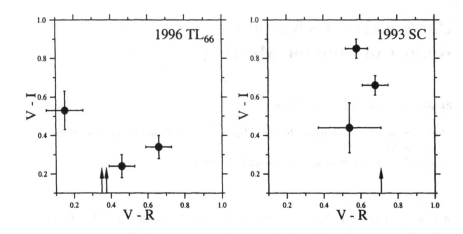

Fig. 1. An example of some reported colours for a couple of 'well-observed' Kuiper Belt Objects. Are the colours actually changing or are there inherent problems with the photometry techniques ?

2   Data Reduction Strategy

After bias correction and flat-fielding (using the twilight sky, dome or median filter night-sky flats), frames can be reduced individually or co-added. If the latter approach is taken, each frame set must be registered with respect to the motion of the target, and separately, the background stars so that the point spread functions (PSF) are (nominally) the same for aperture photometry of the object and the comparison stars.

The key to maximising signal to noise is to select an aperture for photometry which is sufficiently large to sample most of the target object PSF but as small as possible to minimise the sky contribution. There were some differences of opinion as to where the best trade-off lay. Very small apertures would have very little sky contamination but do not sample the whole PSF and are therefore particularly susceptible to small changes in the seeing. However large apertures suffer from a higher sky contamination, and return larger 'formal' errors from the data reduction package.

Definition of the local sky level is critical since this is the dominant noise source. Short term changes in the sky brightness and seeing, make subtraction of the actual background from measurements in frames taken a few hours apart unreliable. However, such frames do provide the opportunity to identify any faint sources which may be present and the appropriate frames can be discarded. Definition of the sky is performed using modal or clipped mean values within a concentric annulus or a selected nearby region free of stars or galaxies.

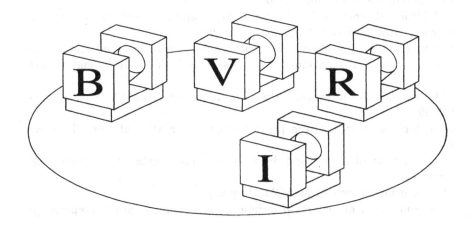

Fig. 2. The VLT solution to simultaneous colour photometry? (Figure drawn by N. McBride following an idea and sketch by Olivier Hainaut and Karen Meech.

3   Photometric Calibration

Photometric calibration is performed using short exposures of standard stars which sometimes need to be de-focused to prevent saturation. The PSF of the standard stars will therefore be rather different from the target images which have longer exposures which 'average out' the seeing (and autoguider jitter). Aperture corrections (to correct for the 'missing' signal outside the aperture) need to be measured or calculated from the PSF of brighter field stars in the target exposures. These may be half a magnitude or more if very small apertures are used. The range of CCD and filter combinations used on large telescopes mean that transformation coefficients are required to place the data into a 'standard' photometric system.

4   Where to Next?

Although similar strategies have been adopted by all observers, their detailed application varied, particularly with the choice of aperture size, and thus with the assessment of aperture corrections and their associated errors. It is clear that the observed range of photometric results could be explained by i) real variations in the magnitudes and colours of the targets, ii) systematic under-estimation of the errors by all observers, iii) a consequence of the different reduction techniques applied by each observer.

It is effect iii) that can be directly tested. Any reduction process should yield the same magnitude or colour if the same target and standard frames are reduced. Thus the following plan was devised:

1) Addition of synthetic KBOs to real data frames — hence the correct answer is known precisely! (O. Hainaut, Jan 99).

2) Observing groups receive data with calibration images and apply their standard reduction techniques. Results sent to N. McBride by end of May 99.

3) Investigation of filter/CCD transformation coefficients (A. Fitzsimmons, Feb 99).

4) N. McBride and S. Green prepare abstract for ACM at Cornell (March 99).

5) Comparison of results and draft presentation circulated to contributors (June 99).

6) Conclusions presented at ACM (July 99).

7) Journal paper on reduction techniques and results of study prepared (in 1999).

5   Note added in proof: What happened next?

The study described above was conducted by M.A. Barucci, C. Delahodde, S. Green, O. Hainaut, N. McBride and J. Romon and reported at "Asteroids Comets Meteors 1999" at Cornell University in July 1999. Delahodde et al. (Delahodde, C.E. & Hainaut, O.R., Discrepancies in TNO colours: A blind test for photometry techniques. ACM 1999 Abstracts Volume, p139, 1999) reported the results of the blind test which involved aperture photometry of a series of synthetic images of objects with a range of brightness and apparent motion. The key results were: i) all observers obtained very similar results for bright objects; ii) the differences for fainter objects were more significant, but those using custom techniques were in general agreement; iii) those using automatic measurement software obtained highly discrepant results for fainter targets which were well outside the formal reported errors from the software; iv) all observers found it difficult or impossible to determine accurate results for faint trailed images.

McBride et al. (McBride, N., Green, S.F., Hainaut, O. & Delahodde, C., Kuiper Belt photometry: Getting the best from your observations. ACM 1999 Abstracts Volume, p136, 1999) reported the results of data reduction simulations to define the optimum aperture as a function of seeing and object motion. The key results were: i) real PSFs are not Gaussian resulting in significant errors in aperture corrections if this assumption is made; ii) the optimum strategy is as follows: 1) determine the seeing FWHM when observing, 2) ensure that the exposure time is selected such that trailing is no larger than 0.3FWHM, 3) set an aperture diameter of 1.3FWHM and apply aperture corrections derived from field stars measured on the same frame. A full report of this study is in preparation for journal publication.

Physical Observations of 1996 TO$_{66}$*

Catherine E. Delahodde[1], Olivier R. Hainaut[1], Hermann Böhnhardt[1], Elisabetta Dotto[2], M. Antonietta Barucci[3], Richard M. West[4], and Karen J. Meech[5]

[1] European Southern Observatory, Alonso de Cordova 3107, Casilla 19001, Vitacura, Santiago, Chile
[2] Osservatorio Astronomico di Padova, Universita de Padova, Vicolo dell'Osservatorio 5, 35122 Padova, Italy
[3] Observatoire de Paris, DESPA, 5, Place Jules Janssens, 92190 Meudon, France
[4] European Southern Observatory, Karl Schwarzschild Straße, 2, 85748 Garching bei München, Germany
[5] Institute for Astronomy, University of Hawai'i, 2680 Woodlawn Drive, Honolulu, HI 96822, U.S.A.

Abstract. TNO 1996 TO$_{66}$ was observed at the ESO NTT during 7 nights, in two groups of 2 and 5 consecutive nights in August and October 1997 (resp). The magnitude of the object displays significant magnitude variations (0.12 mag full amplitude), with a period of 6.25 hrs. The phased lightcurve shows two slightly a-symmetric peaks. Color indexes were also measured, and all the R filter images were stacked and searched for a faint coma.

1 Introduction and Observations

This paper will only give a very short summary of the 1996 TO$_{66}$ observations and preliminary results; a detailed report of the full analysis can be found in Hainaut et al (2000).

We obtained images of 1996 TO$_{66}$ with the SuSI-1 camera on ESO 3.6m New Technology Telescope at La Silla on 1997 August 27 (8 exposures) and 28 (4 exposures), through the Bessel V filter. On October 21–25, 5 nights were partially devoted to our pencil-beam search described in another paper (Böhnhardt et al., in these proceedings). This time, using the EMMI-RILD imager on the same telescope with a set of Bessel filters, we obtained ~ 30 images of a field containing 1996 TO$_{66}$. While most of the images were obtained in the R filter, we also obtained B, V and I images. Most of the fifth night was devoted to obtaining spectra of the TNO.

All the images were processed in the standard way, and the magnitude of the object was measured in a small aperture (3–4"), the sky being estimated in a surrounding, larger annulus. Many (~ 40) field stars were also measured, in order to be used as secondary, in-field standards. The photometric

---

* Based on observations collected at the European Southern Observatory, La Silla, Chile

transformation was calibrated using various standard fields measured during the photometric nights; these transformations were used to estimate the magnitudes of the secondary stars, which, in turn, allowed us to obtain an accurate value of the object's magnitude, even during the nights which were not photometric.

2   Results

The V magnitudes from the August run and the R ones from October show some variations that are significantly larger than the combination of the calibration errors with the photon noise, suggesting that they were caused by a rotation effect. A period searching algorithm was used to confirm this; we used a program derived from that of Harris and Lupishko (1989). A period of 6.250h successfully fits the data from both runs (i.e., it was not necessary to shift the data from one run with respect to the other to achieve the fit). Neglecting the aspect variation effects, this leads to an uncertainty of the period of ± 0.003h. Assuming that the lightcurve amplitude (0.12 mag full amplitude) is entirely caused by the shape of the object, we derive that the elongation of the object is larger than 10%.

The average R magnitude of 1996 TO_{66} was converted into an estimate of its radius of 340$\pm$10km, assuming an albedo of 0.04.

The color indices measured for 1996 TO_{66} are listed in Table 1. With these colors, TO_{66} is among the bluest known TNOs (its $B - V$ index is even bluer than that of Chiron).

All the R frames from the October run were registered on 1996 TO_{66} and stacked. A photometric profile of the object was extracted and compared to a mean stellar profile obtained from a stack of the same images registered on the stars. Both profiles perfectly match down to the 29^{th}mag/sq.arcsec level, indicating that if the object is surrounded by a coma, it must be extremely weak.

3   Discussion

The lightcurve of 1996 TO_{66} presented in this paper is the first TNO lightcurve that has both a complete coverage and a signal/noise ratio sufficient for permitting an accurate rotational study (cf. Davies, 2000, for a review), of course with the exception of Pluto, whose rotation period has been known for years (it should be noted, however, that Pluto's rotation being fully controlled by Charon, it does not give us any information on the rotation rates of the typical TNOs). The period we obtained falls perfectly in the range of what was expected when comparing to other objects suspected to come from the Kuiper Belt (5.9h for (2060) Chiron, 9.98h for (5145) Pholus (Buie and Bus, 1992), 8.87h for (8405) 1995 GO (Brown and Luu, 1997)).

The diameter derived for 1996 TO$_{66}$ makes it the biggest of the TNOs (again, with the exception of Pluto and Charon), and it is also the bluest TNO, with neutral-to-blue colors. These colors suggest a fresh icy surface (as opposed to a surface covered with a crust of slowly photo-damaged organic molecules, which would appear red). This raises the question of the nature of the phenomena that would renew the surface: the most popular explanations being collisions (although it would be difficult to explain the collision rate needed to maintain the surface so blue), and some level of cometary activity. 1996 TO$_{66}$ is therefore a good candidate for coma search. It should also be noted that is is possible that the 0.04 albedo used for the size estimate is actually too low compared to its real value: large, icy bodies tend to have a larger albedo. If proven, this would lead to a smaller radius for TO$_{66}$.

Table 1. Colors of 1996 TO$_{66}$

$B - V$	$V - R$	$R - I$
0.63±0.07	0.40±0.07	0.36±0.07

References

1. Brown , W.R. and Luu, J. X. (1998) CCD Photometry of the Centaur 1995 GO. Icarus **126**, 218–224
2. Böhnhardt, H., Hainaut, O. R., Delahodde, C. E., West, R. M., Meech, K. J., Marsden, B. J. (2000) A Pencil-Beam Search for Distant TNOs at the ESO NTT. In these Proceedings
3. Buie, M. W. and Bus, S. J. (1992) Physical Observations of (5145) Pholus. Icarus **100**, 288–294
4. Davies, J.K. (2000) Physical Characteristics of Trans-Neptunian Objects and Centaurs. In these Proceedings.
5. Hainaut, O. R., Delahodde, C. E., Böhnhardt, H., Meech, K. J., Dotto, E., Barucci, M. A., West, R. M. (2000) Physical Observations of TNO 1996 TO$_{66}$, A&A 356, p.1076-1088.
6. Harris, A. W. and Lupishko, D. F. (1989) Photometric Lightcurve Observation and Reduction Techniques, in Asteroids II, Binzel et al. Eds, p.39–53

TNO Color Survey with the VLT: Pilot Observations with the Science Verification Camera*

Olivier R. Hainaut[1], Hermann Böhnhardt[1], Richard M. West[2], Catherine E. Delahodde[1], and Karen J. Meech[3]

[1] European Southern Observatory, Alonso de Cordova 3107, Casilla 19001, Vitacura, Santiago, Chile
[2] European Southern Observatory, Karl Schwarzschild Straße, 2, 85748 Garching bei München, Germany
[3] Institute for Astronomy, University of Hawai'i, 2680 Woodlawn Drive, Honolulu, HI 96822, U.S.A.

Abstract. In the framework of the Science Verification of the first 8.2-m telescope of ESO's Very Large Telescope, several TNOs were observed in order to measure their magnitude and colors. We present and discuss these results.

1 Introduction and Observations

The Science Verification (SV) Program of ESO's Very Large Telescope (VLT) Unit Telescope 1 took place between August 17 and 31, 1998. Several observation programs were performed, covering a broad range of astronomical topics, in order to verify the suitability of the telescope for astronomical observations. The data, including those on which this paper is based, were released on October 1, 1998 (ESO 1998).

One of the SV programs was aimed at obtaining magnitudes and colors of a set of TNOs. Indeed, a significant fraction of our knowledge of the TNOs' physical properties is based on these parameters (Davies 2000). The typical magnitude range of the TNOs ($R \sim 20$–25) put these objects at the faint limit of the objects for which accurate photometry can be obtained on a 2–4m-class telescope; indeed, fairly long exposures (typically 15–20min), under good sky conditions, are required to achieve a signal-to-noise ratio suitable for color studies (i.e. accurate to a few percents). Moreover, with such telescopes, only a few objects are observable each night. With an 8.2-m telescope, like the VLT Unit Telescopes, the exposure times can be reduced and/or the S/N increased, allowing many objects to be observed during one night. A complete survey of all the minor bodies in the outer Solar System could be performed in just a few nights, resulting in a uniform database of object colors.

---

* Based on observations collected at the European Southern Observatory, Paranal, Chile (VLT-UT1 Science Verification Program)

The SV programs were performed by the Paranal astronomers (Science Verification Team). They selected the observations that were most suited for the weather and seeing conditions. The TNO program was designed as a "back-up", to take advantage of the periods of poor seeing (although the median seeing of Paranal is $\sim 0.65''$, it is not uncommon to have a seeing of up to $\sim 2''$ during some winter nights). Although no time was originally assigned to this program, a total of 3.4h was finally spent on the TNOs, with a seeing varying between 0.7 and $2''$.

The observations were performed using the Science Verification Camera on the VLT UT1. This camera is an imager involving three internal reflections, designed to produce a clean pupil. The detector is a 2k×2k Tektronics CCD (TK2048EB-1522BR07-01) of technical grade. It suffers from a large (~ 300pix) blemish close to the center of the chip, whose sensitivity is very different of the rest of the CCD, and which proved to be hard to correct for by means of traditional flat-fielding. The observers took care not to place the objects in that region. The detector was read in a 2×2 bin read-out, resulting in a pixel size of $0.09''$.

The objects were pre-selected for the quality of their orbit. The actual selection was performed by the observers, based upon the observability of the targets at the time of the observations. Table 1 lists the objects and various related parameters. The observations were obtained through the standard Bessel filters, in a sequence $R\ B\ R\ V\ R\ I\ R$. The repetition of the R filter allows us to monitor the transparency of the sky during the observations, and to disentangle the effects of the rotation (change of the cross-section) from genuine color effects.

The photometric calibration was obtained by measuring various photometric fields. The whole data reduction was handled by the Science Verification Team: they performed the standard bias subtraction and flat-field correction, and computed the photometric transformation for each night and each filter. The images relative to two objects were not released because of their bad image quality.

Using the calibrated images, the TNOs were then identified by blinking the different exposures of the same objects. 1996 KV1, whose motion was very slow at the time of the observations, could not be identified, even after comparing the frames with images of the same field obtained several days later. It is suspected that the object was in front of one of the many stars crowding the field.

The object instrumental magnitudes were measured using the MAGNITUDE/CIRCLE command of the MIDAS package, using a diaphragm of $4''$ diameter. The sky was measured in a ring centered on the object, whose radii were 10 and $15''$. The sky value is obtained by taking a σ-clipped mean of the level in that region. The instrumental magnitudes were corrected of the instrumental and atmospheric effects using the calibration relation provided by the SV Team. No correction was performed to take into account the light that would escape the $4''$ diaphragm.

Table 1. Observation Log

Object	Date [UT]	R[AU]	Δ[AU]	Mag.	Seeing	Exp.Time	Notes
1993 RO	1998-08-24	31.4	30.6	23.1	0.6-1.0	600s	
1994 TB	1998-08-26				1.-2.7	60-100s	1
1996 KV1	1998-08-29	40.8	40.5	23.8	0.9-1.5	300s	2
1996 TL$_{66}$	1998-08-24	35.1	34.7	20.6	0.6-1.0	60-100s	
1996 TO$_{66}$	1996-08-26				>2	60-100s	1
1996 TP$_{66}$	1998-08-24	26.4	25.8	20.9	0.6-0.7	60-100s	

Date refers to the UT epoch of the observations; R and Δ are the helio- and geocentric distances to the object. Mag refers to the expected magnitude of the object (using the information from the MPC web server). Seeing is a measurement of the FWHM images (in arcsec). Exp.Time lists the exposure times that were selected by the Science Verification Team. Notes: 1. Observations not released; 2. Object not found.

2   Results

Table 2 lists the magnitudes and colors obtained. The R magnitudes were converted into absolute magnitudes, and into radii using an albedo of 0.04.

Table 2. Magnitudes, colors and radii

Object	B	V	R	I	$B-V$	$V-R$	$R-I$	Rad
1993 RO		24.37	23.63	23.15		0.74	0.48	52
		±0.03	±0.03	±0.03				±2
1996 TL$_{66}$	21.81	21.46	20.83	20.46	0.35	0.63	0.37	237
	±0.04	±0.04	±0.04	±0.04				±9
1996 TP$_{66}$	22.89	21.86	21.27	20.62	1.03	0.58	0.66	108
	±0.04	±0.04	±0.04	±0.04				±4
Calib. error	±0.02	±0.02	±0.01	±0.02				
Ext. error			±0.06					

The ± listed in the table correspond to the photon noise from the object and the sky in the diaphragm. Calib. error refer to the error listed on the photometric calibration provided by the SV team, and Ext. error is the RMS of the 4 R measurements of each object. Rad is the radius [km] of the object derived from the R magnitude, assuming an albedo of 0.04.

It should be noted that the colors obtained for 1996 TL_{66} present some discrepancies with respect to the values published by others (cf, for instance, Davies 1998). Part of these discrepancies are explained by the fairly large error caused by the very short integration time (60 sec) used for these images. While it is also possible that the measurement method introduced some systematic effects in the magnitude obtained (cf. the splinter session discussion, Green and McBride, these proceedings), these would probably affect all filters in a fairly similar way, and therefore should not affect the colors in a very large way.

Nevertheless, while two of the objects present rather blue-to-neutral colors (that of 1996 TP_{66} being in agreement with other published values), the third one, 1993 RO, presents quite a red $V - R$ index. If confirmed, this object would be one of the reddest TNOs known to date.

The magnitudes obtained from the 4 R images on each objects are constant within the error bars, indicating no rotational effect, and confirming the photometric stability of the sky during the observations.

The R images for each object were registered on the position of the TNO and co-added. A photometric profile was then extracted and compared to that built from various field stars. The TNO profiles do not show any excess compared to the stars, indicating no resolved coma. It should nevertheless be noted that the surface brightnesses reached (25.5 for 1996 TL_{66} and 1996 TP_{66}, 28 for 1993 RO) were not very constraining because of the relatively short exposure times involved. For instance, Delahodde et al. (1998) reached more constraining surface brightnesses for 1996 TO_{66} on a 3.6-m, integrating longer.

3   Conclusion and Projects

This short program proved the suitability of the VLT for TNO studies: colors were measured on 3 objects, using short exposure times, with poor seeing conditions. It should nevertheless be noted that the S/N obtained in 60sec exposures was not optimal: if a full scale survey is launched, longer exposures will be selected. Assuming 300s as a typical typical exposure time, accurate 4-band colors would be measured in 1/2h per object. Four VLT nights would be sufficient to perform a complete, uniform survey of all the TNOs known to date. Observations with FORS (VLT spectro-imager in the visible range) could be combined with additional data from ISAAC (spectro-imager in the near-infrared), to cover a broader spectral range. We can therefore expect to have a complete TNO taxonomy by the next MBOSS meeting.

References

1. Davies, J. K. (2000) Physical Characteristics of Trans-Neptunian Objects and Centaurs. In these proceedings.

2. Delahodde, C. E., Hainaut, O. R., Böhnhardt, H., Dotto, E., Barucci, M., West, R. M., Meech, K. (2000) Physical Observations of TNO 1996 TO_{66}. In these proceedings
3. ESO VLT-UT1 Science Verification (1998), available at http://www.eso.org

Colours of Distant Solar System Bodies

Claes-Ingvar Lagerkvist[1], Mats Dahlgren[1], Andreas Ekholm[1], Johan Lagerros[1,2], Magnus Lundström[1], Per Magnusson[1], and Johan Warell[1]

[1] Astronomical Observatory, Box 515, 751 20 Uppsala, Sweden
[2] Five College Radio Astronomy Observatory, LGRC-617, University of Massachusetts, Amherst, MA 01003 USA

Abstract. The NTT at La Silla, Chile and NOT on La Palma were used for VRI photometry of 9 Edgeworth-Kuiper objects and 3 Centaurs. For the NOT observations growth-curve fitting was applied in the photometric reductions. The spread in colour of the observed Centaurs and Edgeworth-Kuiper objects is large.

1 Introduction

The first object in this population was discovered 1992 by Jewitt and Luu (1993) and up to now (November 1998) more than 70 objects are known. In addition to this there are 8 Centaurs. A review of the physical characteristics of these objects may be found elsewhere in this volume, as well as several papers reporting results from physical observations. One result from the conference was that there are large differences between colours reported by different observers.

2 Observations and Data Reductions

The observations were obtained August 15–17 1996 and January 5–10 1997 with the 2.6-m Nordic Optical Telescope (NOT) on La Palma. During the August observing session, the Brorfelde TEK1024 CCD with a resolution of 0.176 arcsec/pixel (field of view 3.0 arcmin) was used. In January the equipment was the ALFOSC 2k Loral CCD, with a resolution of 0.189 arcsec/pixel (field of view 6.5 arcmin). Details about the observations with the NTT may be found in Magnusson et al. (1998).

The reductions were made with a technique introduced by Howell (1989), using growth curves. The idea is to define a standard growth curve for each frame by measuring the flux of one or a few bright stars on the frame within a series of increasingly larger apertures centered on the star. The flux of the faint source of interest can then be measured using an optimum aperture radius for which the S/N is as high as possible, without concern for whether or not all light is included. The standard growth curve is then used to reduce this value to what it would be for a larger aperture that includes all the light from the source. A detailed description of the reductions will be published elsewhere.

Table 1. Average colors of the observed objects

Object	$V - R$	$R - I$	$V - I$
KBOs			
1993 SB	0.31 ± 0.15	0.66 ± 0.15	1.00 ± 0.15
1994 JR$_1$	0.44 ± 0.15	0.34 ± 0.18	0.77 ± 0.18
	0.75 ± 0.09^b		
	0.36 ± 0.08^c		
1994 TB	0.54 ± 0.15	0.80 ± 0.15	1.36 ± 0.15
	0.85 ± 0.15^a	0.65 ± 0.15^a	
	0.68 ± 0.06^c		
1994 VK$_8$	0.73 ± 0.14	0.77 ± 0.14	1.51 ± 0.16
1995 DA$_2$	0.67 ± 0.11	0.14 ± 0.12^f	0.77 ± 0.12
	0.55 ± 0.11^b	0.50 ± 0.16^b	
1995 DC$_2$	0.52 ± 0.16	1.05 ± 0.20	1.62 ± 0.20
	0.77 ± 0.16^b	0.58 ± 0.16^b	
1995 QZ$_9$	0.38 ± 0.34	0.64 ± 0.23	1.04 ± 0.33
Centaurs			
2060 Chiron	0.38 ± 0.03	0.37 ± 0.03	0.75 ± 0.03
	0.37 ± 0.03^d	0.39 ± 0.04^d	0.76 ± 0.02^d
	0.35 ± 0.02^a	0.54 ± 0.10^a	
5145 Pholus	0.81 ± 0.06	0.78 ± 0.06	1.59 ± 0.06
	0.75 ± 0.02^d	0.84 ± 0.03^d	1.59 ± 0.02^d
	0.84 ± 0.07^a		
	0.78 ± 0.04^c		
8405	0.66 ± 0.14	0.40 ± 0.14	1.03 ± 0.14
	0.41 ± 0.02^d	0.55 ± 0.04^d	0.96 ± 0.03^d
	0.47 ± 0.04^c		
	0.73 ± 0.04^e		

[a]Luu and Jewitt (1996)
[b]Green et al. (1997)
[c]Tegler and Romanishin (1998)
[d]Davies et al. (1998); $R - I$ from their VRI colors
[e]Brown and Luu (1997)
[f]More uncertain value than the error suggests;

3   Results

In Table 1 we present the result from the NOT observations, the NTT observations were reported by Magnusson et al. (1998). In the table comparisons are made with earlier observations. For the results presented at this conference we refer to the other papers in this volume. In Figure 1 we present our result in graphical form. We have here also included the result from the NTT observations. The object 1995 DC_2 was observed with both telescopes and the difference in $V - I$ was substantial. In Figure 1 we give the mean value.

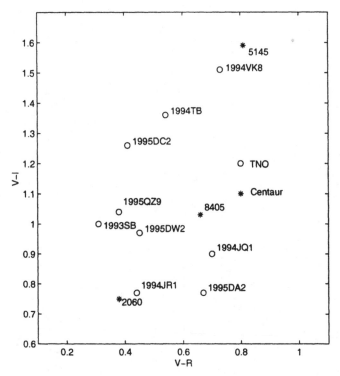

Fig. 1. V-I versus V-R for nine TNOs and three Centaurs

4 Acknowledgement

The observations were carried under ESO proposals 54F.-0618 and 55F.-0827 and the reductions were made as a part of an undergraduate thesis by A. Ekholm.

References

1. Brown, W.R., Luu, J.X. (1997) *Icarus* **126** 218.
2. Davies, J.K., McBride, N., Ellison, S.L. Green, S.F., Ballantyne, D.R. (1998) *Icarus* **134** 213.
3. Green, S.F., McBride, N., Ceallaigh, D.P.O., Fitzsimmons, A., Williams, I.P., Irwin, M.J. (1997) *Mon. Not. R. Astron. Soc.* **290**, 186.
4. Howell, S.B. (1989) *Pub. Astron. Soc. Pac.* **101**, 616.
5. Jewitt, D.C., Luu, J.X. (1993), *Nature* **362**, 730.
6. Luu, J.X., Jewitt, D.C. (1996) *Astron. J.* **112**, 2310.
7. Magnusson, P., Lagerkvist, C.-I., Lagerros, J.S.V., Dahlgren, M., Lundström, M. (1998) *Astron. Nachr.* **319**, 251.
8. Tegler, S.C., Romanishin, W. (1998) *Nature* **392**, 49.

Fig. ...

Acknowledgement

...

References

1. Thomas, W.J. ...
2. ...
3. Green, S.F. ...
4. ...
5. ...
6. ...
7. ...
8. ...

Comet Size Distributions and Distant Activity

Karen J. Meech[1], Olivier R. Hainaut[2], and Brian G. Marsden[3]

[1] Institute for Astronomy, 2680 Woodlawn Drive, Honolulu HI 96822, USA
[2] ESO, Alonso de Cordova 3107, Vitacura, Casilla 19001, Santiago 19, Chile
[3] Harvard-Smithsonian Center for Astrophysics, 60 Garden St., Cambridge, MA 02138, USA

Abstract. We present the results of observations of distant comet nuclei as observed with the Keck II telescope during 1997 December. Our sample included 17 SP Jupiter-family comets, 3 Halley-family comets, and 1 dynamically new comet. The nucleus radii ranged between 0.6 and 12.7 km (assuming a 4% albedo), the average near $R_N \sim 3$ km showing that, in general, the comet nuclei are relatively small. This doubles the known sample of size estimates for the comet population. These data are compared to the size distributions for the Centaurs and the Edgeworth-Kuiper Belt objects.

1 Introduction

The earliest stages of collapse of our solar nebula are not subject to direct observational constraints. However, comet nucleus size distributions are of great interest because they preserve a record of the outer nebula mass distributions in the late stages of planetary formation, as well as a record of collisional evolution. The rate of proto-planetary growth and scattering as a function of heliocentric distance depended on the size and mass distribution of the km-size planetesimals that have survived as today's comets, their surface density in the nebula and their velocity distributions. The estimated sizes of the Edgeworth-Kuiper Belt or Disk objects (EKO) are large compared to known short-period (SP) comet nuclei, although the statistics are still small for both populations.

The SP comets may be collisional fragments from the Edgeworth-Kuiper Belt population which have been injected into resonances to become the SP comets [5]. Bodies larger than 50-100 km probably retain their primordial size distributions [6]. As shown by Stern [20], the present Edgeworth-Kuiper disk is probably collisionally very active – especially for smaller objects likely to evolve into SP comets. The larger disk bodies may reflect the scale of instabilities in the outer solar nebula, whereas the long-period comets (LP) that have been stored in the Oort cloud may not have been subjected to collisions, so that their size distributions may be primordial.

The size distribution of the EKOs is a critical boundary condition for understanding the formation of the solar nebula. Likewise, the size distributions of the Centaurs and the SP and LP comets will be important. However, interpreting the results will be difficult because of a large number of observational biases.

2   Observing Program

Observations were obtained on 1997 December 28 and 29 on the Keck II telescope using the Low Resolution Imaging Spectrometer (LRIS) in its imaging mode. The CCD was read out with 2 amplifiers with gains of 1.97 and 2.10 e^- ADU^{-1} and a read noise of 6.3 and 6.6 e^-. The images were well sampled with a pixel scale of 0.215 $''$ pix^{-1} and seeing ranging between 0.5$''$ and 1.0$''$, FWHM. Both nights were photometric. Images were taken through the V and R filters on the Johnson system, and were guided at non-sidereal rates. Comets were observed over $3 < r < 24$ AU to make estimates of their nucleus sizes. For a detailed discusison of the specifics of the data and reductions, see Meech et al. [19].

3   Size Determination Issues

Comet nucleus sizes are currently reported in the literature from 3 observational techniques: (*i*) the coma subtraction method used by Lamy et al. [14], (*ii*) simultaneous optical and infra-red observations, and (*iii*) observations of distant, inactive comet nucleii. None of these techniques is a direct measurement – all have underlying assumptions and are model dependent. However, in most cases there is reasonable agreement between nucleus size estimates for comets measured by more than one technique (see Table 1).

Table 1. Nucleus Size Measurement Comparison

Comet	R_N Infrared	p_v	R_N Dist. Obs.	p_v	R_N Coma Sub.	p_v	Ref
2P/Encke	2.5±0.5	0.08	2.8-6.4	0.04			[8,11]
22P/Kopff			2.46	0.04	3.3-3.8	0.04	[15], here
28P/Neujmin 1	10.0±0.5	0.025	10.42	0.04			[3], here
29P/SW1	20±2.5	0.13			15.4±0.2	0.04	[4,18]
9P/Tempel 1			2.10	0.04	3.9×2.8	0.04	[13], here
10P/Tempel 2	5.9±0.4	0.02	3.07, 8×4×4	0.04			[1,12], here
55P/Tempel-Tuttle	1.8±0.4	0.06	1.8±0.2	0.04			[9,10]
81P/Wild 2	3.0±0.3	0.02	2.0±0.04	0.04			[9,17]
46P/Wirtanen			0.7	0.04	0.6±0.02	0.04	[2,13]

3.1 Coma subtraction

Lamy et al. [14] uses the Hubble Space Telescope (HST) to image the inner coma of active comets, and models the coma contribution to determine the flux from the nucleus and infer a nucleus size. The coma brightness, $B(\rho)$, where ρ is the projected distance from the nucleus in arcsec, is modeled by a function:

$$B(\rho) = [k_c/\rho + k_n \delta(\rho)] \otimes PSF \tag{1}$$

Here k_c and k_n are normalization constants, δ is the delta-function of the nucleus, and PSF is the point spread function of the telescope. The brightness is modelled and compared to the observed surface brightness profiles to determine the nucleus contribution. The assumptions inherent in this technique which can give rise to systematic effects are numerous. First, many comae are not symmetric and may not be well modelled by a radially symmetric profile. In addition, this method assumes that the coma profile intensity drops as a function of ρ^{-1} which is true only for a steady-state coma unaffected by radiation pressure. Finally, the technique assumes a geometric albedo, $p_v = 0.04$, and a dust phase function, $\beta = 0.04$ mag deg^{-1}. The phase function in particular, has only been measured on a few comets between 5-30° (Meech and Jewitt [16]) and may not be applicable to the high phase angles at which the comets are often observed with HST.

3.2 Infrared observations

Simultaneous optical and thermal infrared comet observations can give an estimate of both the instantaneous nucleus size and geometric albedo. The technique utilizes the following relations for the optical and thermal fluxes, F:

$$F_{opt} \propto R_N^2 p_v \phi(\alpha) \tag{2}$$

$$F_{thermal} \propto \epsilon R_N^2 \phi_{thermal}(\alpha) \tag{3}$$

However, just as in the previous case, assumptions must be made about the optical phase function, ϕ, as well as the thermal phase function, $\phi_{thermal}$. Measurements of $\phi_{thermal}$ have not been made for comets, therefore the thermal phase function measured for asteroids is used: $\phi_{thermal} = 0.005 - 0.017$ mag deg^{-1} (for $\alpha < 30°$). Because infrared detectors are not as sensitive as optical detectors, often the comet must be fairly close to the sun for a detection, which can imply a large phase angle, α, leading to greater uncertainty in these terms. The close proximity to the sun can also result in significant coma around the nucleus, and thus some sort of model must be used to account for the thermal and optical signal from this coma. Finally, a thermal model (e.g. the standard thermal model or the isothermal latitude model, etc.) must be used to interpret the radiometry, and this creates even more uncertainty in the final results.

3.3 Distant comet photometry

The scattered light, m_N, from the nucleus of an inactive comet is related to the size of the nucleus, R_N, and the geometric albedo, p_v, by:

$$p_v R_N^2 = 2.235 \times 10^{22} r^2 \Delta^2 10^{0.4(m_\odot - m_N)} / \phi(\alpha) \qquad (4)$$

Here the heliocentric and geocentric distances, r and Δ [AU], and the solar magnitude, $m_\odot$, are known, and the only values which must be assumed are the albedo and the phase function $\phi(\alpha)$. Because the comets must be at large distances to be inactive, and are therefore very faint, typically they are observed near opposition to get the maximum observing time on the object, and therefore the phase angles, α, are very small. Thus any errors in assuming a phase function of $\phi(\alpha) = 0.035$ mag deg^{-1}, which is common, are also very small.

The main uncertainty in this technique is ascertaining whether the nucleus is really inactive or if there is some residual dust coma which may be unresolved or at too low a surface brightness to be detected. Low-level activity can be ruled out if the comet has been observed over a range of r and the reduced magnitude, $m(1, 1, 0)$, has remained constant (with possible rotational modulations). This, unfortunately, requires observations over a long time base (*e.g.* years).

¿From the above discussion, it is clear that all of the methods of comet nucleus size determination have assumptions and/or depend on models, so that none are direct measurements. We have a direct measurement only for 1 comet which was visited by spacecraft: 1P/Halley. Nevertheless, it appears that the distant comet photometry technique is the least prone to assumptions.

4   Discussion and Results

In a study of the sizes of SP comets which have appeared in the literature, Fernández et al. [7] have found that most comets have absolute nuclear brightnesses lying in the range of 15-19 mag, implying radii between 0.5-3.3 km for $p_v = 0.04$. However, given that the present work has found that there is often coma seen on comets at very large r, *i.e.* well beyond $r = 6$ AU where water-ice sublimation is a strong driver of activity, observations must be made at even larger r to ensure that the radius estimates are not contaminated by coma. For example, a bare comet nucleus with $m(1, 1, 0) = 19$ will reach m ≈ 25 near $r = 4.5$ AU, so beyond this distance observations to determine the true distributions of nucleus sizes will require very large aperture telescopes.

This work has shown that observations from 1-2 nights of large telescope time can significantly increase the number of nucleus size estimates. The data presented here have increased the number of sizes to 36 – a 50% increase. The nucleus radii range over $0.6 < R_N < 12.7$ km, with the most common size between 1-3 km. How many nucleus sizes are needed to make valid statistical comparisons between the different dynamical classes? Figure 1 shows a comparison of the size distributions for the observed SP comets, the LP and dynamically new (DN) comets, the EKOs and the Centaurs. There is a clear difference in the size distributions between the SP comets and the EKOs, but there are many observational biases unaccounted for. Because of the difficulty in obtaining direct SP nucleus measurements owing to their faintness, the sample is very incomplete, especially for small sizes. This incompleteness is even more severe for the EKOs. Also, in order to interpret the comparison of the distributions in terms of the early solar system, we must understand the collisional evolution of the SP comets, the dynamical transport mechanisms into the inner solar system, and the evolutionary effects on the SP comets (*e.g.* sublimation and splitting). This leaves us with the important question – to what extent will the understanding of the SP comet sizes help towards understanding the small EKO distribution?

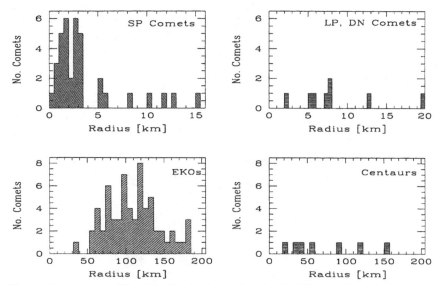

Fig. 1. Comparison of the size distributions of comets in different dynamical classes.

References

1. A'Hearn, M. F., Campins, H., Schleicher, D. G., Millis, R. (1989): *ApJ* **347**, 1155-1166
2. Boehnhardt, H. R., West, R. M., Babion, J. (1997): *A&A* **320**, 642-651
3. Campins, H., A'Hearn, M. F., McFadden, L.-A. (1987): *ApJ* **316**, 847-857

4. Cruikshank, D. P., Brown, R. H. (1983): *Icarus* **56**, 377-380
5. Davis, D. R., Farinella, P. (1997): *Icarus* **125**, 50-60
6. Farinella, P., Davis, D. R. (1996): *Science* **273**, 938-941
7. Fernández, J. A., Tancredi, G., Licandro, J. (1996): *Rev. Mexicana* **4**, 112
8. Fernández, Y. R. *et al.* (1998): *BAAS* **30**, 42.06
9. Fernández, Y. R. *et al.* (1999): *IAU Colloq. 168* in press
10. Hainaut, O. R., Meech, K. J., Boehnhardt, H., West, R. M. (1998): *A&A* **333**, 746-752
11. Jewitt, D., Meech, K. (1987): *AJ* **93**, 1542-1548
12. Jewitt, D., Luu, J. (1989): *AJ* **97**, 1766-1790
13. Lamy, P. L. (1998): *IAUC 7000*
14. Lamy, P. L., *et al.* (1996a): *Icarus* **119**, 370-384
15. Lamy, P. L., A'Hearn, M. F., Toth, I., Weaver, H. A. (1996b): *BAAS* **28**, 8.04
16. Meech, K. J., Jewitt, D. C. (1987): *A&A* **187**, 585-593
17. Meech, K. J., Newburn, R. L. (1998): *BAAS* **30**, 42.03
18. Meech, K. J., *et al.* (1993): *AJ* **106**, 1222-1236
19. Meech, K. J., Hainaut, O. R., Marsden, B. G. (1999): in prep.
20. Stern, S. A. (1996): *A&A* **310**, 999-1010

Ion Irradiation of Minor Bodies in the Outer Solar System

Giovanni Strazzulla

Osservatorio Astrofisico, Città Universitaria, I-95125 Catania, Italy

Abstract. Fluxes of charged particles impinge on matter forming small objects in the outer Solar System during its evolution from (pre)solar grains to planetesimals to the final object, and produce a number of effects whose knowledge might be relevant for understanding the evolution of those objects. This research is based on laboratory simulations of relevant targets bombarded with charged particles. Here I discuss experimental results obtained, in recent years, on physico-chemical effects induced on frozen gases and mixtures simulating relevant targets.

1 Introduction

The knowledge of the physico-chemical properties of the surfaces of the small objects in the external Solar System is very scarce. Key objectives are to understand which materials and processing mechanisms produce the observed colors and/or spectra. Materials could reflect, at least partially, the composition of pre-existing (interstellar, IS) grains. IS-grains could be "reprocessed" in the solar nebula and we have to investigate to which extent energetic processing (UV and/or energetic particle irradiation) might modify pre-existing grains and/or grains that are reformed in the solar nebula and accrete to form larger bodies. The complexity of the questions is such that it is particularly relevant to approach the problem(s) from several different points of view: ground based and space observations are obviously needed as well as theoretical and laboratory work.

My group has been involved for about 20 years in a laboratory research on the effects induced by energetic ions on relevant materials (ices, silicates, hydrocarbons [1]). Here I present some recent results and try to outline the relevance this laboratory investigation might have to understand the processing of objects in the outer Solar System.

2 An Organic Crust?

The composition of minor bodies in the outer Solar System could reflect, at least partially, the composition of pre-existing (interstellar, IS) grains. UV photons and cosmic ions deposit a large amount of energy on grains in the interstellar medium. Thus, based on experimental results obtained in several laboratories, we can expect, starting from simple ices, the production of many new species including organic refractory materials [2]. In the solar nebula these grains are, to some extent, "reprocessed" and volatiles in the forming object are due to accretion from a gas whose composition has been quenched at high temperature or by direct accretion of the precursor IS grains after losing the very volatile species [3]. The irradiation history, during accretion, may then be at least as important as the one in the ISM. This is particularly true if we assume that accretion occurred during an active phase of the young Sun (T-Tauri phase). Once formed, these objects have been exposed for a long time to the flux of galactic cosmic rays. A review of the effects expected on Oort cloud comets has been presented [4]. In particular it has been estimated that the external (0.1-0.5 m) layers of a comet were subjected to an irradiation dose of 600 eV/molecule. Deeper layers were subjected to a lower dose because the most abundant but less energetic ions are stopped by the external layers. Although that dose is less than those suffered in the early evolutionary phases, its role could be of great relevance for cometary physics. It has been, in fact, suggested that a comet exposed to background particle radiation obtains an outer web of non-volatile material which will lead to the formation of a substantial "crust" [4]. The existing observational data on the colors of cometary nuclei, Halley's in particular, and of some small bodies in the external Solar System are compatible both with the hypothesis that they have a substantial organic crust and that only the very outer cometary skin is dark. In the latter case darkness might be due to an organic rich layer or to a very porous silicate skin.

How thick is the crust (if it exists)? Strazzulla and Johnson [4] estimated a thickness of 0.1-0.5 m depending on the actual density of the external layers of the object. That estimation is based on the experimentally measured cross section ($\simeq 10^{-17}$ cm^2) for the conversion of frozen methane to a refractory residue [5, 6]. We have now measured the fraction of carbon incorporated in the refractory material for more cases and showed that the cross section of the process has values 4×10^{-17} cm^2 for methane, 4×10^{-16} cm^2 for benzene and 3×10^{-14} cm^2 for polystyrene [7]. These results indicate that whatever is the hydrocarbon, it is converted to a refractory, insoluble residue. The cross section of the process depends on the molecular weight of the original material (it is about 10^3 times higher for polystyrene than for methane). This implies that if the original material going to form comets already contains organics with high molecular weight (as it could be the case for organics already present on interstellar pre-cometary grains) the crust formation process is much more efficient than in the case previously considered [4]. Thus the crust could be much more thick (up to a factor 10^3) than the 0.1-0.5 m estimated in [4].

Other experiments demonstrated that the organic crust has been already formed during bombardment at low temperature [8]. This gives credit to the hypothesis that cometary crust can be already formed during the long stay far from the Sun and its development does not requires a first passage (heating) in the inner Solar System. This last result is relevant for the trans-Neptunian objects that could be a reservoir of short period comets.

It is also relevant that the IR spectra (in the 3.4 μm region) of some residues obtained from frozen methane and butane reproduce very well those observed in the diffuse interstellar medium and in the organic residue from Murchison meteorite [9]. The progressive carbonization of irradiated targets influences, of course, their optical properties also in the UV-Vis range. In particular, the slope of the reflectance spectra (400-800 nm) changes with ion fluence. First extracted residues (at moderate doses) show "red" spectra becoming flatter and flatter under further ion irradiation [10].

It has been suggested that there are significant compositional grouping of comets apparently related to place of formation [11]. The majority of comets originating in the Kuiper belt (and formed beyond Neptune and Pluto) appear in fact to be significantly depleted in the carbon-chain molecules whose abundance could be related to the amount of irradiation. Comets of the Halley family, originally formed from the regions of Uranus and Neptune and then expelled to the Oort's cloud, exhibit a "normal" composition.

A'Hearn et al. [11] suggest that the difference is primordial, due to a different composition of pre-cometary material. They discard the hypothesis that the long residence of comets in the Oort cloud, outside the heliosphere, has allowed to galactic cosmic rays to provide a second source of carbon-chain molecules. I agree with this conclusion, also in view of the recent measurements by the Voyager and Pioneer interplanetary spacecrafts, now moving in the Kuiper belt region, that have confirmed the radially increasing intensity of an additional radiation source [12]. The surface of Kuiper's comets should be even more irradiated than the one of the Oort's comets. However, in my opinion, the observed variability is not necessarily connected to a difference in the initial composition of pre-cometary material but related to the formation process itself. If we suppose that Oort's comets, are formed, in the Uranus and Neptune region, by homogeneous accretion, we expect, as discussed in [1], an homogeneous exposure to ion irradiation. Kuiper's comets, formed beyond Neptune and Pluto, should, in this scenario, be formed by accretion of different planetesimals thus being expected to be strongly dishomogeneous. In fact even among trans-Neptunian objects different colors have been measured but this could also be due to the presence on their surfaces of different low-volatility materials that would be lost as they eventually pass near the Sun.

3   Effects on Specific Species

As said above we have little knowledge of the actual composition of trans-Neptunian objects. Thus it is extremely difficult to say if there are specific molecules whose presence clearly indicate the origin and/or processing of those materials. In particular we have not yet found, in the laboratory, any specific molecule, whose detection may clearly indicate that ion processing has been responsible for its formation. All of the parent molecules identified on comets are also observed on icy grains in dense molecular clouds where they could be formed by direct condensation from the gas phase (e.g., CO), or by chemical reactions on grain surfaces (e.g. CO_2) or by energetic processing (e.g., CO and CO_2 produced by ion irradiation of water/methanol mixtures).

In addition to the dominant nitrogen and to solid methane CO and CO_2 have been detected on Triton's surface [13]. Pluto has some CO, but less than Triton has, and no CO_2 [14, see also De Bergh et al. these proceedings]. We are investigating the possibility that CO and/or CO_2 are produced in situ by cosmic ion irradiation of appropriate frozen ices. The energetic ions in Neptune's magnetic field might impact the surface and form CO_2 from CO. Pluto's surface is impacted by cosmic rays and the solar wind, but it does not have a nearby source of a strong magnetic field like Triton has. To this end we are measuring CO/CO_2 molecular number ratio obtained after ion irradiation of a number of relevant frozen mixtures.

In a recent paper a new spectrum of the Centaur 5145 Pholus has been presented [15]. The authors matched the observed spectrum with five components among which 15% of water ice that accounts for observed features at about 1.6 and 2 μm and 15% methanol ice that accounts for the observed feature at about 2.27 μm. As outlined by the same authors methanol has an additional band at about 2.33 μm that does not seem to be observed on Pholus.

A large number of experiments indicate that ion irradiation, among other effects, induces not only the formation of different molecules, both less and more volatile, but also changes the shape and relative intensity of bands due to the same molecule. In particular we have irradiated thin frozen methanol films and studied its IR spectrum in the 3-17 μm region [16]. Now we are irradiating more thick films, in order to study the spectral region between 2 and 3 μm. The first results indicate that after irradiation the 2.33 μm band become much weaker than the 2.27 μm one. This could help to explain its lack in the observed spectrum. It is however quite possible that the 2.27 μm band is due to a different species and, in particular, CN bearing species seems to be promising.

Recently, we have performed new experiments irradiating ternary mixtures H_2O:NH_3(or N_2):CH_4. It has been shown that among the newly synthesized species there is a R-O-C$\equiv$N group (R is a group in the organic residue) which exhibits a band at about 4.62 μm [17]. This band remains after warming up to room temperature and furnish the best available fit to the so called X-C$\equiv$N band observed towards some young stellar sources still embedded in their placental cloud [18]. Up to now we have not been able to detect any overtone of that band that could be expected at about 2.3 μm.

References

1. Strazzulla G. (1997) Ion bombardment of comets, in From Stardust to Planetesimals Y.J. Pendleton, A.G.G.M. Tielens eds., ASP Conf Series Book, S. Francisco, p. 423-433.
2. Jenniskens P., Baratta G.A., Kouchi, G.A., de Groot, M., Greenberg, J.M., Strazzulla, G. (1993) Carbon dust formation on interstellar grains A&A **273**, 583-600.
3. Yamamoto T. (1991) Chemical theories on the origin of comets, in Comets in the Post-Halley Era, R. Jr Newburn, M. Neugebauer, J. Rahe eds., Kluwer, Dordrecht, pp. 361-376.
4. Strazzulla G. and Johnson R.E. (1991) Irradiation effects on comets and cometary debris, in Comets in the Post-Halley Era R. Jr Newburn, M. Neugebauer, J. Rahe eds., Kluwer, Dordrecht, p. 243-275.
5. Foti G., Calcagno L., Sheng K.L. and Strazzulla G. (1984) Micrometer-sized polymer layers synthesized by MeV ions impinging on frozen methane, Nature **310**, 126-128.
6. Strazzulla G., Calcagno L. and Foti G. (1984) Build up of carbonaceous material by fast protons on Pluto and Triton. A&A **140**, 441-444.
7. Strazzulla G. (1999) Ion irradiation and the origin of cometary materials. In Proc. of the workshop "The Origin and Composition of Cometary Materials" Altwegg K. et al. eds, Kluwer Acad. Publ., submitted
8. Strazzulla G. and Baratta G.A. (1992) Carbonaceous material by ion-irradiation in space, A&A **266**, 434-438.
9. Pendleton Y.J., Sandford S.A., Allamandola L.J., Tielens A.G.G.M. and Sellgren K. (1994) Near-infrared absorption spectroscopy of interstellar hydrocarbon grains, ApJ **437**, 683-696.
10. Andronico, G., Baratta, G.A., Spinella, F., Strazzulla, G. (1987) Optical evolution of laboratory-produced organics: applications to Phoebe, Iapetus and outer belt Asteroids, A&A **184**, 333-336.
11. A'Hearn, M.F., Millis, R.L., Schleicher, D.G., Osip, D.J., Birch, P.V. 1995, Icarus, 118, 223
12. Cooper, J.F., Christian, E.R., Johnson, R.E. 1996, 31st Sci Assembly of COSPAR, The Univ. of Birmingham, England, p.71 (abstract)
13. Cruikshank D.P., Roush T.L. et al. (1993) Ices on the surface of Triton Science **261**, 742-745.
14. Owen T.C., Roush T.L. et al. (1993) Surface ices and atmospheric composition of Pluto, Science **261**, 745-748.
15. Cruikshank D.P., Roush T.L. et al. (1999) The composition of Centaur 5145 Pholus Icarus, in press.
16. Palumbo M.E., Castorina A.C., Strazzulla G., (1998) Ion irradiation effects on frozen methanol (CH_3OH) A&A, in press.
17. Palumbo M.E., Strazzulla G., Pendleton Y.J., Tielens A.G.G.M., (1999) R-O-C≡N species produced by ion irradiation of ice mixtures: comparison with astronomical observation, ApJ , submitted.
18. Pendleton Y.J., Tielens A.G.G.M. Tokunaga A.T., Bernstein M.P. (1999) The interstellar 4.62 μm band, ApJ , in press.

Surveys of the Distant Solar System

Alan Fitzsimmons

Astrophysics and Planetary Science Division, The Queen's University of Belfast, Belfast BT7 1NN, Northern Ireland

Abstract. The rapid development in the 1990's of surveys for distant Centaurs and Trans-Neptunian Objects (TNOs) is reviewed. With 94 TNOs and 9 Centaurs found by the end of 1998 (ignoring Pluto in this instance), plus another 6 TNOs observed on one night only, the apparent sky density of these objects at $21 \geq m_R \geq 26$ is now well measured, with there being approximately one TNO brighter than $m_R = 23.0$ per square degree of sky. However, at present the bright and faint ends of the magnitude distribution are poorly constrained, and there is no strong evidence for or against a single power-law applying to TNOs of masses from Pluto down to 1P/Halley. The non-discovery of any TNOs in-$situ$ at heliocentric distances $R_h > 50$ AU is found to be significant. This could be caused by a real 'edge' to the Kuiper-belt, or due to a change in the population as a whole at these distances preventing detection.

1 Introduction

The obvious method for discovering and identifying distant Solar system bodies is through their orbital motion about the Sun (in this paper I define the outer Solar system as being beyond Jupiter at $R_h = 5.2$ AU). The range of possible orbital elements can mean that unless some assumptions are made concerning the orbit, relating the apparent motion over a short arc to the true orbital motion is fraught with difficulty. A major problem is that searches for distant Solar system bodies are generally confined to the ecliptic, and the possibility exists of confusion between truly slow-moving distant objects and main-belt asteroids near their stationary point.

If we assume that the Earth has a circular orbit (a good approximation) as well as the distant object (possibly a bad approximation!), then the rate of motion at opposition can be easily shown to be

$$\omega = \frac{147.8}{R_h + \sqrt{R_h}} \ \text{arcsec hour}^{-1} \tag{1}$$

Here R_h is the heliocentric distance of the body and 147.8 is the mean orbital angular motion of the Earth in arcsec hour^{-1}. Because of this simple mapping, any discovered objects can then have a distance quickly estimated. Hence the majority of surveys of distant Solar system objects have been performed near opposition [2].

Note that another way of avoiding main-belt contamination would be to search at quadrature $i.e.$ along the Earths orbital motion vector, as used with

HST observations [5]. In this case the apparent motion of the object is simply due to the component of its orbital motion in the plane of sky, again allowing its distance to be immediately estimated in the case of a circular orbit via

$$\omega = \frac{147.8}{R_h^{3/2}} \;\; \text{arcsec hour}^{-1} \tag{2}$$

2   Pre-1992 Surveys

2.1   Photographic surveys

The first directed large-scale surveys of the outer Solar system were photographic through necessity, as photographic plates were the only medium available for wide-field imaging. Generally these plates were searched for distant objects either through scanning for short trails, or more frequently through blinking plate pairs obtained hours or even days apart. In the modern era of software analysis, one can only hold in awe the diligence and expertise of those researchers engaged in this work. It is unlikely that this will ever be be repeated again; today Schmidt plates and films are processed using automatic measuring machines such as SuperCOSMOS.

These first photographic surveys were however fruitful. Tombaugh [28] visually blinked 1530 square degrees of the ecliptic down to $m_{pg} = 18$, and discovered Pluto, the first Trans-Neptunian Object (TNO) of any kind (ignoring long-period comets). A later search by Kowal [21] discovered the first Centaur (2060) Chiron. In order to achieve this, 6400 square degrees were searched down to $m_R = 18.5$. The most recent large-scale photographic search reported was that by Luu & Jewitt [23], who surveyed 297 sq. degrees down to $m_R = 19.0$, but were unsuccessful in detecting any new distant objects.

2.2   CCD surveys

From these surveys, it was obvious that the space density of small bodies beyond Jupiter was extremely small at the magnitudes reachable photographically. Luckily the advent of CCD detectors in astronomy in the 1980's brought with them a detector that was both linear with received flux up to saturation, and more importantly gave a high quantum efficiency. The drawback was the small area coverage on sky, indeed in the late 1990s we are only just approaching mosaics of detectors capable of imaging a tenth of the area of a large Schmidt plate.

The small area of the early CCDs did not prevent attempts to detect distant minor bodies. Luu and Jewitt [23] searched 0.34 sq. deg. down to $m_R = 24.0$, while Levison and Duncan [22] searched 4.9 sq. deg. to a limiting magnitude of $m_R = 21.8$. Both teams used R-band filters as darker sky at V is outweighed by the higher quantum efficiency at R to give a higher SNR at red wavelengths. These searches and some later ones used a manual blinking technique. Such a technique is severely inefficient for modern large CCD arrays, and software techniques have now become the norm [4]. Neither search found any distant objects, although comparison with the next section shows that both teams may be considered unlucky in not being successful.

3   1992 QB₁ and Beyond

After a sixty-two year gap, Pluto was finally given company beyond Neptune with the discovery by Jewitt and Luu [17] of the first recognised TNO, 1992 QB$_1$, at $R_h = 40.9$ AU. The following year saw another five TNOs discovered, and the pace has since accelerated. Jewitt and Luu [18] surveyed 1.2 sq. deg. down to $m_R = 24.8$ and discovered 7 TNOs. Williams et al. [30] surveyed a smaller area of 0.7 sq. deg down to only $m_R = 22.5$ and discovered two more. Irwin et al. [14] also surveyed 0.7 sq. deg down to a fainter magnitude of $m_R = 22.5$ and found another 2 TNOs. Continuing their work, Jewitt et al. [2] surveyed 8.3 sq. deg. to $m_R = 24.2$ discovering 15 TNOs and 2 Centaurs, and Jewitt et al. (1998) report surveying a huge 51.5 sq. deg. to $m_R = 22.5$ to find another 13 TNOs. Finally Brown et al. [3] searched 12 sq. deg. but were unsuccessful in confirming any new objects brighter than 21.

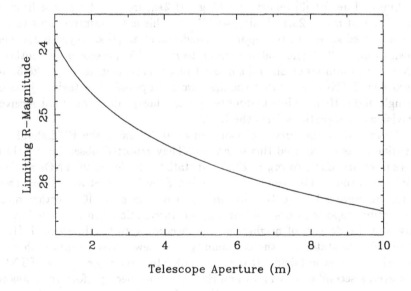

Fig. 1. Detection limits of point sources in a 600-second exposure at R for 1-m to 10-m class telescopes, assuming the same operating characteristics of the Wide-Field Camera on the 2.5-m Isaac Newton Telescope on La Palma.

Apart from the last search, these CCD surveys all used 2-m class tele-scopes to integrate on sky for typically between 5 and 15 minutes. This is a natural limit to the integration time, for in 15 minutes a TNO at 30 AU will move $\sim$ 1 arcsec, and the dark sky at international observatories will result in a background level of $\sim 10^4$ electrons (depending on pixel size, filter, phase of moon etc). Hence a longer exposure time will not significantly improve the SNR of a TNO because of trailing. The 3-σ detection limit for a point source in a 600-second R-band integration at a good site with 1 arcsec seeing is shown in Figure 1. Combined with extra sources of noise such as background sources, cosmic rays etc, this gives a natural limiting magnitude to such 'classical' 2-m searches of $m_R \simeq 24$ with modern CCDs.

In order to overcome this limit on such instruments, several searches have used a 'shift and add' technique. Here the same field is continuously imaged, and the images are then re-stacked in some way to account for the probable motion of any TNOs. In its simplest application, this will result in trailed galaxies and stars, but any TNOs will appear as point sources if the apparent motion was correctly calculated. Such co-addition of the individual frames generally also results in the removal of cosmic rays in the final image. Further refinements allow for the subtraction of background stars and galaxies before co-addition of the frames. In this way a small area of sky is probed to a significantly fainter limiting magnitude, similar to the 'pencil-beam' surveys of high-redshift galaxies used in cosmology.

Luu and Jewitt [3] report searching 101.2 sq. arcmin. down to a limiting magnitude of $m_R = 26.1$, finding six TNOs. This is the faintest limit to any ground-based survey so far reported. Gladman et al. [1] surveyed 0.3 sq. deg. down to $m_R = 24.6$ and 180 sq. arcmin. to $m_R = 25.6$, discovering another 5 TNOs. Fitzsimmons et al. [10] surveyed just 0.17 sq. deg. to $m_R = 25.5$ and discovered 2 TNOs. Note that the first two surveys only saw their objects on a single night. Hence while undoubtedly real, those objects have not received provisional designations form the IAU.

At this point the survey of Cochran et al. [5] using the HST should be mentioned, as it too used this technique. They reported observing a field of $\sim$ 4 sq. arcmin. down to $m_R = 27.9$, and statistically found 29 TNOs at these faint magnitudes. This result was questioned by Brown et al. [2], but these criticisms were subsequently answered by Cochran et al. [6]. Further more, the number reported does not now appear unrealistic when compared to the observed sky density of brighter objects according to Gladman et al. [1]. It is only fair to state that the community still views these results with some suspicion, as a second field also observed failed to show any evidence of TNOs. However, a second observing run on HST has since been performed to answer these queries; it makes no sense to the author to further pursue this matter until these new data have been analysed.

3.1 Centaurs

The same surveys are also sensitive to detecting objects closer than Neptune. To date 9 Centaur objects have been discovered with orbits lying between

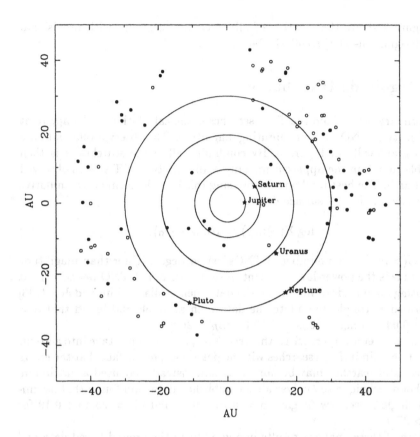

Fig. 2. Position of all known TNOs and Centaurs on December 31st 1998. Open circles designate objects with poorly constrained orbits, dark circles designate bodies observed on two or more apparitions. The orbits and positions of the outer gas giants are also marked.

Jupiter and Neptune, four of them ((2060) Chiron, 1994 TA, 1995 DW$_2$ and 1998 QM$_{107}$) by 'classical' 2-m surveys as listed above [21,2]. The remaining 5 Centaurs ((5145) Pholus, (7066) Nessus, (8405), 1997 CU$_{26}$ and 1998 SG$_{35}$) were all discovered by the Spacewatch facility [16]. While the primary remit of Spacewatch is the observation of Near-Earth Objects, its automated moving object detection software allows the detection of all objects with measurable apparent motion. The 'classical' 2-m surveys result in a sky surface density of ~ 0.5 per sq. deg. at $m_R \geq 24.2$ [2], while detailed modelling of the Spacewatch data by Jedicke and Herron [16] gives a power law of $\alpha = 0.61^{+0.70}_{-0.40}$. Obviously a significant increase in the number of Centaurs known is required before the relative populations of TNOs and Centaurs is firmly established and the source region can be unequivocally unidentified.

Figure 2 shows the location of all TNOs and centaurs with provisional IAU designations at the end of 1998.

4 Magnitude Distribution

The primary result from the TNO searches is the calculation of the apparent sky density of TNOs to some limiting magnitude. The observational data are shown graphically in Figure 3. By combining all of the searches it is then possible to obtain the apparent magnitude distribution of TNOs as observed from Earth. The best method of expressing this is in the form of a cumulative power law for the sky surface density.

$$\log \Sigma(> m_R) = \alpha(m_R - m_0) \tag{3}$$

Here $\Sigma(> m_R)$ is the number of TNOs per sq. deg. brighter than magnitude m_R and α is the power-law exponent, meaning there is 1 TNO per sq. deg. at a limiting magnitude of m_0. Using the data then available, Luu and Jewitt [3] performed a straight line fit to the data then available and found that $\alpha = 0.54 \pm 0.04$ and $m_0 = 23.20 \pm 0.10$ for $m_R > 26.6$.

Gladman et al. [1] criticised this technique, as it does not take into account either upper limits from searches with no detections, or the fact that the errors in successful searches may be non-Gaussian. Instead they used a maximum-likelihood technique to analyse all the published data, and found that assuming a single power-law fit gave $\alpha = 0.76 \pm 0.11$ and $m_0 = 23.40 \pm 0.19$ for $m_R > 27.9$.

Both of these analyses results in a good fit to the ground-based data and are close to each other. Of course, this agreement should be expected as both are attempting to fit a straight-line through similar data! However the important difference arises at the faint end of the distribution, as the first fit is not compatible with the HST result, while the second is. It is clear that more 'deep' searches are now required at $m_R \leq 26$ to fully constrain the magnitude distribution at size approaching those of comets. Indeed, another simplification is the assumption of a *single* power-law. The collisionally evolved asteroid-belt does not exhibit a single size distribution. As shown by the studies of collisional evolution in the Kuiper-belt by Davis and Farinella [7], it is likely that this is the case for TNOs as well.

In an interesting aside, Gladman et al. [1] show that the ratio of objects detected in 'deep' and 'classical' surveys is given by

$$\frac{N_d}{N_c} = \frac{1}{6} 10^{1.6\alpha} \tag{4}$$

Thus 'deep' surveys are best for the overall magnitude distribution in any magnitude range, as one will simply discover more objects. Unfortunately, such surveys can only give limited information on the orbital characteristics

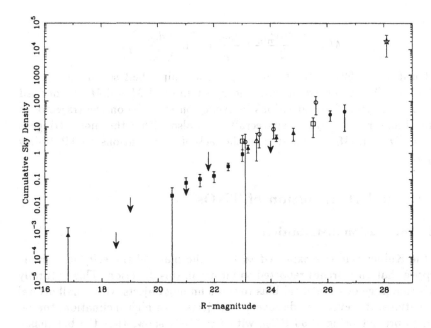

Fig. 3. Observed magnitude distribution of TNOs. Dark symbols refer to observations by Jewitt & Luu, open symbols by other observers. The star refers to the deep HST survey[5].

of the TNOs found. To investigate dynamics via orbital element distributions, astrometry of TNOs must be obtained at a minimum of two oppositions for even rough orbit determination. Hence 'classical' surveys are still necessary, as only objects observable in a single exposure are likely to be followed in subsequent years.

5 Mass Constraints

The apparent magnitude distribution leads naturally to an estimate for the total number larger than a minimum size, if one assumes an albedo for all objects. This is normally taken to be $p_R \simeq 0.04$ through comparison with cometary nuclei (Meech et al., this volume), where the subscript R denotes the albedo at red wavelengths near 6500Å. Indeed the first measurements of the albedos of TNOs (Thomas et al., this volume) support such low reflectivities. If this is the case, then the observed objects have diameters d between 100 km and 800 km and Jewitt et al. [1] then calculate that $N(d > 100\text{km}) \simeq 7 \times 10^4$. However care must be taken, as Pluto has $p_R \simeq 0.7$ [12], while (2060) Chiron has $p_R \simeq 0.14$ [4]. Luu and Jewitt [3] show that for a value $\alpha = 4.0$, the total mass contained within minimum and maximum diameters d_{min} and d_{max} is

$$M = \frac{\pi \rho N d_{min}^3}{2} \left(\frac{0.04}{p_R} \right)^{3/2} \ln \left(\frac{d_{max}}{d_{min}} \right) \tag{5}$$

For $d_{min} = 100\,\text{km}$, the total mass of the Kuiper-belt so far observed is $\sim 0.1 M_\oplus$. This is consistent with the upper limit of $M \leq 5 M_\oplus$ as derived from the analysis of perturbations by Anderson et al. [1] on the trajectories of the Pioneer and Voyager spacecraft. It is also within the more stringent limit of $M \leq 1.3 M_\oplus$ imposed from the lack of perturbations on 1P/Halley [13][31].

6   Orbital Distribution of TNOs

6.1   Inclination distribution

As the Kuiper-belt was expected to lie in the plane of the ecliptic, it is no surprise that all searches reported so far are in this direction. This naturally leads to a large observational bias towards finding objects with small orbital inclinations. However, the discovery of objects with high inclination, the record at present being 1996 RQ_{20} with $i = 31.6°$, shows that the belt has a substantial thickness. Modelling by Jewitt et al. [2] found a true inclination distribution with FWHM of 30° could reproduce the observed distribution. If our eyes were sensitive enough, the densest part of the Kuiper-belt would be seen to cover roughly one-quarter of the sky.

6.2   Radial distribution

Observations of proto-planetary disks such as those around β-Pictoris and Fomalhaut) show structures extending for 100's of AU's from their parent star. While the rate of accretion, and hence the size of bodies, will vary strongly as a function of heliocentric distance, there is no *a-priori* reason to doubt the existence of TNOs as far as $R_h = 100\,\text{AU}$ or more. Indeed, dynamical modelling by Duncan et al. [9] and Stern [27] has demonstrated that the inner Kuiper-belt at $R_h < 45\,\text{AU}$ has been significantly eroded by gravitational perturbations from Neptune. This is effectively the region so far explored by ground-based telescopes, and implies that we should find a significantly higher space density of TNOs beginning at $R_h \simeq 50\,\text{AU}$.

Figure 4 shows the heliocentric distances of TNOs at discovery. At first glance, it appears that the number density of TNOs actually falls rapidly beyond $\sim 45\,\text{AU}$. However, there is a strong observational bias in this figure. The central problem is that the brightness of TNOs is effectively proportional to $\sim R_h^{-4}$, so that probing these distant regions with medium sized 2-m class telescopes is extremely difficult. Even so, Figure 2 underlines the point that to date observational searches capable of detecting TNOs at $R_h \geq 50\,\text{AU}$ with diameters $\geq 400\,\text{km}$ have failed to do so. Dones [8] first suggested that the

observed number of faint TNOs was incompatible with the belt extending to
$R_h > 50$ AU. This hypothesis was supported by Jewitt et al. [1], who through
Monte-Carlo techniques found that $\sim 40\%$ of discovered TNOs should be at
$R_h > 50$ AU if the belt extends to 200 AU. Note that the discovery of 1996
TL$_{66}$ with a semi-major axis of $a = 85.2$ AU does not affect this argument,
as it is believed this is the first identified member of a 'scattered disk' popu-
lation [24], different from the unperturbed disk being considered here.

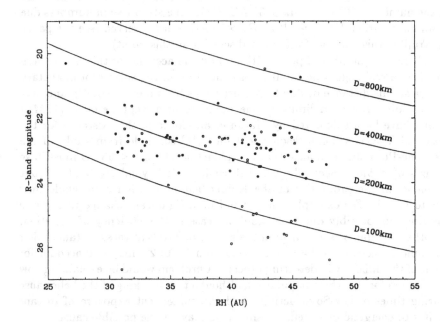

Fig. 4. Position and apparent magnitude of TNOs at the date of discovery. Dark
circles show objects with well-defined orbits, open circles mark objects with have
only been observed at one apparition. Besides 94 TNOs with IAU designations, 6
TNOs seen on one night only during 'deep' searches [3] [1] are also plotted. Lines
of equal diameter are shown for an albedo of 4%

Gladman et al. [1] analysed this possible discrepancy in depth. They
showed that if the number density is given by $n(R_h) \propto R_h^{\beta}$, and the size
distribution is given by $dn(r) = r^{-q}dr$, then the fractional number of TNOs
observed to be more distant than R_h from the sun is given by

$$f = \left[\left(\frac{R_{min}}{R_h} \right)^{\gamma} - 1 \right]^{-1} \qquad (6)$$

where ($\gamma = 5 - 2q - \beta$). Using the total population then discovered, for
$R_{min} = 30$ AU, $q = 4.8$ and $\beta = -2.0$, $f = 4\%$. Hence they concluded that
at that time there was no significant evidence of a drop in numbers at large

distances. However, the choice of taking $R_{min} = 30$ AU is somewhat questionable. The majority of objects discovered within 40 AU are dynamically trapped by Neptune in the 2:3 resonance and are therefore unrepresentative of a smoothly varying disk. If instead one takes $R_{min} = 40$ AU (and even this may not be distant enough), then taking values of q between 4.0 and 4.8 results in the prediction of between 30% to 50% of objects being discovered at $R_h > 50$ AU. As all 52 objects discovered at $R_{min} > 40$ AU lie within 50 AU, one can safely conclude that this model also supports a large decrease in the number of TNOs at $R_h > 50$ AU. Hence the steep rise in numbers due to an unperturbed Kuiper-belt at $R_h > 50$ AU is not observed (see the paper by Trujillo, this volume, for further discussion of this point).

There are a number of possibilities that could account for this. First there could be a real edge, due to either a truncation in the solar protoplanetary disk near this distance or due to perturbations from more distant planetary-mass bodies. However, limits to the existence of a Neptune-mass ($17M_{\oplus}$) planet have been calculated by Maran et al. [26] from the existence for the observed TNOs, while Jackson and Killen [15] conclude from study of the Neptune-Pluto 3:2 resonance that any planet at $53 \leq R_h \leq 63$ AU must have a mass of $< 5M_{\oplus}$. Secondly, the disk could exist beyond $R_h > 50$ AU but a change in the properties of the Kuiper-belt population as a whole may act to hide it. For example, there could be a significant change in the size distribution, possibly due to a rapid decrease in the efficiency of accretion. Another possibility is a change in the albedo of TNOs at these distances. For example, a decrease in the mean albedo from 4% to 2% may well account for some of the difficulty in detecting objects. A problem would be explaining how this occurs for the whole population, although the shrinking of the heliopause during times of low-Solar activity and the consequent exposure of distant TNOs to energetic interstellar cosmic rays may be one possible cause.

7   Conclusions

The 1990's have seen an explosion in the number of observable distant minor bodies. Standing successes include the ability to be able to regularly observe TNOs and to a lesser degree Centaurs, which is helpful when requesting telescope time! The sky surface density is now well measured for $21 \geq m_R \geq 26$. Also the total mass of the observed Kuiper-belt is in line with unrelated constraints. Puzzles currently before us are understanding the probable 'edge' to the Kuiper-belt at $R_h \simeq 50$ AU, knowing the true inclination distribution, and constraining the high-end mass distribution.

All of these problems can be addressed by performing larger-scale surveys than those currently in progress, both in the ecliptic and tens of degrees from it. Figure 5 shows the number of objects discovered per year up until the end of 1998 - within six years, the number of TNOs observed on more than one night (the requirement for a provisional IAU classification) has risen from

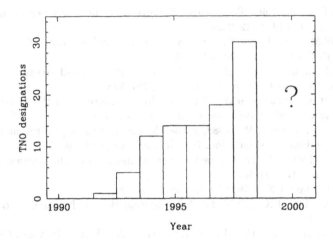

Fig. 5. Number of TNOs receiving provisional designations in the 1990's.

zero to nearly one hundred. Any meeting to discuss work on minor bodies in the outer Solar system may well have between 10 and 100 times more objects to consider than in 1998. I wish the reviewer in that meeting luck.

References

1. Anderson J.D. et al. (1995) Improved Bounds on Nonluminous Matter in Solar Orbit. Astrophys. J. **448**, 885–892
2. Brown M.E. et al. (1997) An Analysis of the Statistics of the Hubble Space Telescope Kuiper Belt Object Search. Astrophys. J. Lett **490**, L119–L122
3. Brown M.J.I., Webster R.L. (1998) A Search for Bright Kuiper Belt Objects. Pub. Astr. Soc. Australia **15**, 176–178
4. Campins H. et al. (1994) The Color Temperature of (2060) Chiron: A Warm and Small Nucleus. Astron. J. **108**, 2318–2322
5. Cochran A. et al. (1995) The Discovery of Halley-Sized Kuiper-Belt Objects using the Hubble Space Telescope. Astrophys. J. **455**, 342–346
6. Cochran A.L. et al. (1998) The Calibration of the Hubble Space Telescope Kuiper Belt Object Search: Setting the Record Straight. Astrophys. J. Lett **503**, L89–L93
7. Davis D.R., Farinella P. (1997) Collisional Evolution of Edgeworth-Kuiper Objects. Icarus **125**, 50–60
8. Dones L. (1997) Origin and Evolution of the Kuiper Belt. In From Stardust to Planetesimals, Eds Y.J. Pendleton and A.G.G.M. Tielens, ASP Conference Series **122**, 347–365
9. Duncan M. et al. (1995) The Dynamical Structure of the Kuiper Belt. Astron J. **110**, 3073–3081
10. Fitzsimmons A. et al. (1999) A Deep Search for Kuiper-Belt Objects. Astron. Astrophys, submitted.

11. Gladman B. et al. (1998) Pencil Beam Surveys for Faint Trans-Neptunian Objects. Astron. J. **116**, 2042–2054
12. Grundy W.M., Fink U. (1996) Synoptic CCD Spectrophotometry of Pluto over the Past 15 Years. Icarus **124**, 329–343
13. Hamid S.E. et al. (1968) Influence of a Comet Belt Beyond Neptune on the Motions of Periodic Comets. Astron. J. **73**, 727–728
14. Irwin M. et al. (1995) A Search for Slow-Moving Objects and the Luminosity Function of the Kuiper-Belt. Astron. J. **110**, 3082–3092
15. Jackson A.A., Killen R.M. (1988) Planet X and the Stability of resonances in the Neptune-Pluto System. Mon. Not. R. astr. Soc. **235**, 593–601
16. Jedicke R., Herron, J. (1997) Observational Constraints on the Centaur Population. Icarus **127**, 494–507
17. Jewitt D.C., Luu J.X. (1993) Nature **362**, 730–732
18. Jewitt D.C., Luu J.X. (1995) The Solar System Beyond Neptune. Astron. J. **109**, 1867–1876
19. Jewitt D. et al. (1996) The Mauna Kea-Cerro-Tololo Kuiper Belt and Centaur Survey. Astron. J. **112**, 1225–1238
20. Jewitt D. et al. (1998) Large Kuiper-Belt Objects: The Mauna Kea 8K CCD Survey. Astron. J. **155**, 2125–2135
21. Kowal C. (1977) Icarus **77**, 118–123
22. Levison H.F., Duncan M.J. (1990) Astron. J. **100**, 1669–1675
23. Luu J.X., Jewitt D.C. (1988) Astron. J. **95**, 1256–1262
24. Luu J. et al. (1997) A New Dynamical Class in the Trans-Neptunian Solar System. Nature **387**, 573–575
25. Luu J.X., Jewitt D.C. (1998) Deep Imaging of the Kuiper Belt with the Keck 10 Meter Telescope. Astrophys. J. Lett. **502**, L91–L94
26. Maran M.D. et al. (1997) Limitations on the Existence of a Tenth Planet. Planet Sp. Sci. **45**, 1037–1043
27. Stern A. (1996) On the Collisional Environment, Accretion Timescales and Architecture of the Massive, Primordial Kuiper Belt. Astron. J. **112**, 1203–1211
28. Tombaugh C. (1961) in Planets and Satellites, Eds. G.P. Kuiper and B.M. Middlehurst, Univ. Chicago Press, Chicago, 12
29. Trujillo C., Jewitt D. (1998) A Semiautomated Sky Survey for Slow-Moving Objects Suitable for a Pluto Express Mission Encounter. Astron. J. **115**, 1680–1687
30. Williams I.P. et al. (1995) The Slow Moving Objects 1993 SB abd 1993 SC. Icarus **116**, 180–185
31. Yeomans D.K. (1986) Physical Interpretations from the Motions of Comets Halley and Giacobini-Zinner. In 20th ESLAB Symposium on the Exploration of Halley's Comet, ESA Publications Division, ESTEC, Noordwijk, **2**, 419–425

Dust Measurements in the Outer Solar System

Eberhard Grün[1], Harald Krüger[1], and Markus Landgraf[1,2]

[1] Max-Planck-Institut für Kernphysik, Heidelberg, Germany
[2] NASA Johnson Space Center, Houston, TX, U.S.A.

Abstract. Dust measurements in the outer solar system are reviewed. Only the plasma wave instrument on board Voyagers 1 and 2 recorded impacts in the Edgeworth-Kuiper belt (EKB). Pioneers 10 and 11 measured a constant dust flux of 10-micron-sized particles out to 20 AU. Dust detectors on board Ulysses and Galileo uniquely identified micron-sized interstellar grains passing through the planetary system. Impacts of interstellar dust grains onto big EKB objects generate at least about a ton per second of micron-sized secondaries that are dispersed by Poynting-Robertson effect and Lorentz force. We conclude that impacts of interstellar particles are also responsible for the loss of dust grains at the inner edge of the EKB. While new dust measurements in the EKB are in an early planning stage, several missions (Cassini and STARDUST) are en route to analyze interstellar dust in much more detail.

1 Introduction

The detection of an infrared excess at main sequence stars started a renewed interest in the outer extensions of our own solar system dust cloud. Especially, the observation of a dust disk around β-Pictoris stimulated comparisons with the zodiacal cloud. Whereas the observed β-Pictoris disk extends from a few 10's to several 100 AU distance from the central star, the zodiacal cloud has been observed only out to a few AU from the Sun [9]. The zodiacal light photometer onboard Pioneer 10 observed zodiacal light out to about 3.3 AU from the Sun where it vanished into the background. The infrared satellites IRAS and COBE observed dust outside the Earth's orbit. Asteroidal dust bands [17] were comprised of two components: a hot ($\sim 300\,K$) component of nearby ($\sim 1\,AU$) dust and a warm ($\sim 200\,K$) component that corresponds to dust in the asteroid belt. Cold dust ($< 100\,K$) corresponds to interstellar dust beyond the solar system near the galactic plane.

Detection of Edgeworth-Kuiper belt objects (EKOs) of up to a few 100 km diameter confirmed the existence of objects outside the planetary region in the distance range where the disk around β-Pictoris has been found. Such objects had been theoretically predicted in order to explain the frequency of occurrence of short period comets and from models of the evolution of a disk of planetesimals in the outer solar system. About 30 objects have been found to date outside about 30 AU from the Sun [11]. Theories predict an extension of kilometer-sized objects out to 3,000 AU from the Sun. Because of mutual collisions and because of their interactions with the environment, generation of dust has been predicted [21] in the Edgeworth- Kuiper belt (EKB).

It is obviously very difficult to recognize faint extensions of a dust cloud when the observer sits inside the dense parts of this cloud, therefore, previous attempts to detect these portions of the zodiacal cloud by astronomical means failed. In the next section we review in-situ spacecraft measurements that pertain to dust in the outer solar system. Ironically, so far, the best evidence for dust in the outer solar system comes from measurements inside Jupiter's distance.

In section 3 we review our knowledge about interstellar dust in the local interstellar medium. After a discussion of dynamical effects of and consequences on dust in the EKB (section 4) we conclude in section 5 by a review of future attempts and plans to get more information about dust in the outer solar system.

2   Spacecraft Observations

Four spacecraft carried dust detectors beyond the asteroid belt: the early Pioneers 10 and 11 [10] and recently the Galileo and Ulysses [4] spaceprobes. Figure 1 shows trajectories of spacecraft beyond Jupiter's orbit. The two Voyager spacecraft did not carry specific dust detectors. However, during the passage of Voyager 2 through the newly discovered G-ring in Saturn's ring system it was recognized that the plasma wave instrument onboard was able

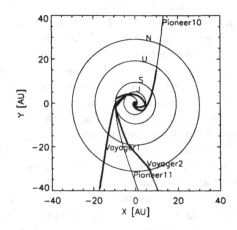

Fig. 1. Dust measurements in the outer solar system. Trajectories of space spacecraft beyond Jupiter's orbit are shown in projection onto the ecliptic plane. Heavy lines indicate regions where dust measurements have been reported by the investigators [10] [8]

to detect impacts of micron-sized dust onto the spacecraft skin [7]. Flybys of Uranus and Neptune confirmed this effect. The investigators attempted to identify dust impacts in the data obtained during occasional tracking of the spacecraft in interplanetary space beyond 6 AU [8]. They found significant impact rates out to about 30 and 50 AU, respectively. The authors state a flux of 10^{-11} g particles of about $5 \cdot 10^{-4}$ m^{-2}s^{-1}. This flux is more than a factor 10 above the corresponding zodiacal dust flux at 1 AU [3]. The problem with the chance dust detector onboard the Voyager spacecraft is (1) that this instrument has never been calibrated for dust detection, i.e. its sensitivity has not been experimentally determined, (2) the sensitive area of the detector has only been derived from theoretical considerations, and (3) the distinction of impact events from noise events has not been verified, i.e. observations of micron-sized dust by Voyager have only been reported from regions of space where no other spacecraft took similar measurements. Only Cassini measurements may confirm the Voyager findings and provide a cross calibration. Therefore, quantitative results from the Voyager observations have to be taken with great caution.

Beginning at about 3 AU from the Sun, measurements of about 10 micron-sized dust by the Pioneer 10 detector (mass threshold $\sim 8 \cdot 10^{-9}$ g) showed a constant dust density out to 18 AU [10]. At this distance the nitrogen in the pressurized dust sensor froze out and prohibited measurements further away from the Sun. Dust measurements by Pioneer 11 (mass threshold $\sim 6 \cdot 10^{-9}$ g) out to Saturn's distance were reported by Humes [10]. During the three passages of Pioneer 11 through the heliocentric distance range from 3.7 to 5 AU the detector observed a roughly constant dust flux. This led Humes to conclude that this dust had to be on highly eccentric orbits that have random inclinations (if the particles are on bound orbits about the Sun).

Interstellar dust grains passing through the planetary system have been detected by the dust detector onboard the Ulysses spacecraft [4]. These observations provided the unique identification of interstellar grains by three characteristics: 1. At Jupiter's distance the grains seemed to move on retrograde trajectories opposite to orbits of most interplanetary grains and the flow direction coincided with that of interstellar gas [20], 2. A constant flux has been observed at all latitudes above the ecliptic plane, while interplanetary dust displays a strong concentration towards the ecliptic [6] [12] and 3. The measured speeds (despite their substantial uncertainties) of the interstellar grains were high (≥ 15 km s^{-1}) which indicated orbits unbound to the solar system, even if one neglects radiation pressure effects [5].

3   Interstellar Dust Characteristics

Clearly identified interstellar grains range from 10^{-15} g to above 10^{-11} g (see Figure 2) with a maximum at about 10^{-13} g. The deficiency of small grain

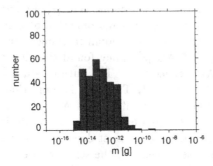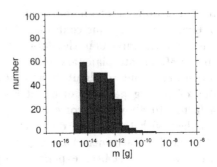

Fig. 2. Mass distributions of interstellar grains observed by the Galileo (left) and Ulysses (right) dust instruments [13]. The detection threshold of the detectors is 10^{-15} g at $26 \, \mathrm{km \, s^{-1}}$

masses is not solely introduced by the detection threshold of the instrument but indicates a depletion of small interstellar grains in the heliosphere. Estimates of the filtering of 0.1 micron-sized and smaller electrically charged grains in the heliospheric bow shock region [2] and in the heliosphere itself [13] show that these small particles are strongly impeded from entering the planetary system by the interaction with the ambient magnetic field.

The mass density of interstellar grains detected by Galileo and Ulysses is displayed in Figure 2. Below about 10^{-13} g it shows a strong deficiency of small grains due to heliospheric filtering. Above about 10^{-12} g a flat distribution is suggested (corresponding to a slope of -4 of a differential size distribution). The total mass density of the observed grains is $7 \cdot 10^{-27} \, \mathrm{g \, cm^{-3}}$. The upper limit ($10^{-9}$ g) of the mass distribution is not well determined: if we extend the flat distribution to bigger masses, about $1.5 \cdot 10^{-27} \, \mathrm{g \, cm^{-3}}$ per logarithmic mass interval has to be added.

Even bigger radar meteor particles ($m \geq 10^{-7}$g) have been found [19] to enter the Earth's atmosphere with speeds above $100 \, \mathrm{km \, s^{-1}}$. These speeds are well above the escape speed from the solar system which confirms their interstellar origin. Big interstellar meteors arrive from a broad range of directions and are not collimated to the interstellar gas direction as smaller particles are. At present the total mass flux of big interstellar meteor particles is not known. Therefore, an extrapolation from the Ulysses observations up to 10^{-7}g has a large uncertainty.

Frisch et al. [2] summarize properties of the local interstellar cloud (LIC, Table 1). If the total hydrogen density is complemented by helium (with a number density ratio $n_{He}/n_H = 0.1$) the total gas mass density in the LIC is about $7 \cdot 10^{-25} \, \mathrm{g \, cm^{-3}}$. The canonical gas-to-dust mass ratio of 100 (from "cosmic abundance" considerations) compares favorably with the observed values. However, several modifications of the dust mass density in the LIC are suggested. Firstly, small "classic" astronomical interstellar grains may

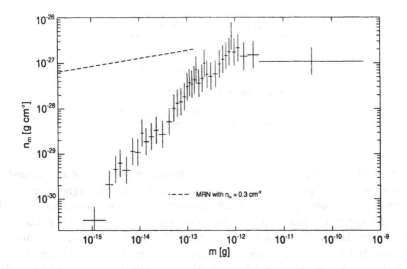

Fig. 3. Mass density of interstellar dust. Measurements by Galileo and Ulysses in the inner heliosphere are compared with "classic" astronomical grains expected to be present in the local interstellar medium as well. The astronomical grains are represented by the MRN distribution [15] corresponding to a total hydrogen density of $0.3\,\mathrm{cm}^{-3}$

need to be added to the interstellar grains detected by Galileo and Ulysses. Mathis, Rumpel and Nordsieck (MRN, [15]) represent these particles by a power law with exponent -3.5 in the radius range from 5 to 250 nm. Figure 3 shows this distribution by a dashed line. A mass density of 5 to $7 \cdot 10^{-27}\,\mathrm{g\,cm}^{-3}$ has to be added if these small particles are present in the LIC - which we will assume in the further discussion. Secondly, supernova shocks passing through the interstellar medium process interstellar grains by shattering and evaporation, i.e. part of the grain material is put into the gas phase and shows up as absorption lines in the spectra of nearby stars. Refractory elements like Mg, Si, Ca, and Fe have been identified. Therefore, the total content of heavy elements in the LIC is further increased. Both the mere existence of big interstellar grains ($> 10^{-13}\,\mathrm{g}$) and the total mass of interstellar grains in the LIC have important consequences for the understanding of the interstellar medium. Big grains couple to the interstellar gas over much longer length scales than the small "classic" interstellar grains, both by friction and by gyro-motion imposed by the interstellar magnetic field. Grains in the diffuse interstellar medium are electrically charged by the competing effects of electron collection from the ambient medium and the photo-effect of the far UV radiation field. The charged dust grains couple to the magnetic field

Table 1. Characteristics of the local interstellar cloud (LIC, after [2]). Both hydrogen number densities (neutral: n_{H0}, and ionized n_{H+}) are given

Item	Value
n_{H0}	$0.22\,\text{cm}^{-3}$
n_{H+}	$0.1\,\text{cm}^{-3}$
Temperature	6,900 K
Magnetic Field	0.15 to 0.6 nT

which itself is strongly coupled to the ionized component of the interstellar medium. A simple comparison shows that the electromagnetic coupling length is several orders of magnitude shorter than the frictional length scale for LIC conditions. However, for particles with masses $> 10^{-12}\,\text{g}$ the gyro radius exceeds the dimension of the LIC and, therefore, these particles are not expected to move with the LIC gas. $10^{-7}\,\text{g}$ particles could travel more than 100 pc through the diffuse interstellar medium (at LIC conditions) with little effect. This mechanism provides the basis for any heterogeneity in the gas-to-dust mass ratio. Locally there may be significant variations in the gas-to-dust mass ratio and hence deviations from the "cosmic abundance" which has to be preserved only on the average over large regions of space.

4   Dust Dynamics in the Edgeworth–Kuiper Belt

Impacts of interstellar grains onto objects in the Edgeworth-Kuiper belt generate dust locally. In order to estimate the amount of dust generated we represent the size distribution of EKOs by a simple power law [18] [11] $n(s)\,ds = N_0\,s^{-4}\,ds$, with $N_0 = 1.3 \cdot 10^{19}$, i.e. 35,000 objects are in the observed EKO size range of $5 \cdot 10^4\,\text{m}$ to $s_{max} = 1.6 \cdot 10^5\,\text{m}$ radius. This distribution has constant mass per equal logarithmic size interval, but most of the cross sectional area is in the smallest objects. Therefore, the size distribution is truncated at $s_{min} = 100\,\text{m}$. By integrating this size distribution over the full size range we arrive at a total cross section of about $4 \cdot 10^{17}\,\text{m}^2$.

Impact experiments with micron-sized projectiles into water ice [1] suggest that at $26\,\text{km\,s}^{-1}$ impact speed about 10^4 times the projectile mass is excavated and ejected mostly in form of small particulates. Because of the low gravity of EKOs we assume that most secondary particles are ejected at speeds in excess of the escape speed of 0.1 to $10\,\text{m\,s}^{-1}$. The interstellar dust mass flux of $2 \cdot 10^{-16}\,\text{g\,m}^{-2}\,\text{s}^{-1}$ generates ejecta particles at a rate of $8 \cdot 10^5\,\text{g\,s}^{-1}$. A more detailed calculation [21] arrives at similar values. About the same amount of micron-sized dust is generated by mutual collisions among EKOs [18], i.e. about 2 tons of dust are generated in the Edgeworth-Kuiper belt every second. This value has a large uncertainty and is probably a lower limit since much cross sectional area could be in smaller EKOs than previously assumed.

In the absence of big planets the Poynting-Robertson effect is the most important dynamical effect on dust in the EKB. The time τ_{pr} (years) for a dust grain of radius s (cm) and density $\rho_d (g\,cm^{-3})$ to spiral to the Sun from a circular orbit at distance r (AU) under the Poynting- Robertson effect can be estimated from

$$\tau_{pr} = 7 \cdot 10^6 \, s \, \frac{\rho_d}{Q_{pr}} r^2. \tag{1}$$

A 10 μm sized particle with $\rho_d = 2.7\,g\,cm^{-3}$ and radiation pressure efficiency $Q_{pr} = 1$ would need about 6.5 Myrs to reach the inner planetary system starting from an initial circular orbit at 50 AU. The radiation pressure efficiency Q_{pr} decreases for particles smaller than the effective wavelength of solar radiation. Liou et al. [14] have shown that during their orbital evolution micron-sized grains are trapped in mean motion resonances with the outer giant planets. The biggest effect comes from resonances with Neptune which prolongs the particle's residence time inside about 40 AU significantly. Therefore, the Poynting-Robertson life time given in eqn. (1) is only a lower limit for the dynamical life time of EKB dust particles. Liou et al. found that in many cases ($\sim 80\,\%$) particles are ejected out of the solar system by passages close to Jupiter before they could reach the region of the inner planets.

There is another force that becomes increasingly important in the outer heliosphere: Lorentz scattering of electrically charged interplanetary grains by solar wind magnetic field fluctuations [16]. Charging of dust particles by the combined solar UV photo-effect and electron capture from solar wind plasma results in a surface potential of about +5 V leading to a charge-to-mass ratio that varies as s^{-2}. Carried out by the solar wind plasma (at speeds of 400 to $800\,km\,s^{-1}$ away from the Sun) the interplanetary magnetic field forms a Parker spiral. The dominant azimuthal component of the magnetic field varies as $1/r$ with heliocentric distance. At 1 AU the Lorentz force is comparable to solar gravitational attraction for particles of $s \sim 0.1$ microns. Therefore, in the EKB at 50 AU, Kepler orbits of even micron-sized particles are strongly affected by the Lorentz force.

Near the solar equatorial plane ($\sim \pm 15°$) the magnetic field changes polarity two to four times per solar rotation period (25.2 d). Above and below this equatorial region (which is roughly centered at the ecliptic plane) a unipolar magnetic field prevails that changes its polarity with the 11-year solar cycle. Both short and long-term magnetic field fluctuations lead to diffusion of grain orbits mostly in inclination [16] but also some outward convection reduces the Poynting-Robertson effect of 10 micron-sized and smaller grains.

This dynamical evolution has to be compared with the collisional life times in the outer solar system. There the dominant flux of micron-sized projectiles is from interstellar grains. Therefore, we calculate the collision rate C_{coll} and the corresponding mean collisional life time $\tau_{coll} = 1/C_{coll}$ for interstellar grains. We follow a similar calculation for interplanetary grain collisions by Grün et al. [3], especially, we use the same collisional parameters for our calculation. Although the EKB is beyond the planetary region it is still located inside the heliosphere and some filtering of interstellar grains may occur. Therefore, we cut-off the interstellar size distribution at about 10^{-15}g.

Figure 4 shows the life times of EKB particles at 50 AU due to collisions with interstellar grains. For comparison we show the pure Poynting-Robertson life times. The life times of 10 micron-sized particles are dominated by collisions with interstellar grains. Considering the prolonged residence time due to Lorentz scattering and mean motion resonances, even smaller EKB grains are destroyed by interstellar dust impacts before they can reach the inner planetary system where they mix-in with zodiacal dust. We conclude that impacts of interstellar particles are not only a major contributor of dust in the EKB but may also be responsible for the loss of dust grains at the inner edge of the EKB. A complication is that the density of the ambient interstellar medium is variable on time scales of 10^5 to 10^6 years and extrapolations from the present state cannot be easily made. Impacts of interstellar grains may play an important role for the existence and structure of extended dust sheets like that around β-Pictoris.

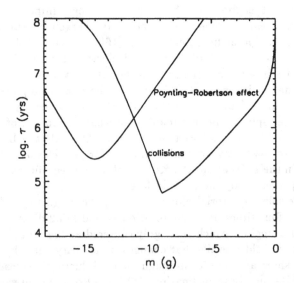

Fig. 4. Life times of dust grains at 50 AU from the Sun. Collision life times are calculated for a flux of interstellar projectiles with masses $> 10^{-15}$ g. For comparison life times due to the Poyting-Robertson effect are shown. The Poyting-Robertson life time is calculated for particles on circular orbits starting at 50 AU distance; no planetary resonance effects have been considered

5   Future Measurements

New measurements of interstellar grains passing through the planetary system are expected from the Cassini and STARDUST missions. Cassini with its Cosmic Dust Analyzer (CDA) was launched in October 1997. It will commence dust measurements at its final Venus flyby in June 1999 and continue to make interplanetary dust measurements until its arrival at Saturn in 2004. The Cassini CDA combines a large area dust detector ($0.1\,\mathrm{m}^2$) with a mass analyzer for impact generated ions. Thereby, the first medium-resolution ($M/\Delta M \sim 20$ to 50) compositional measurements of interstellar grains will be performed.

The STARDUST Discovery mission will collect samples of cometary coma and interstellar dust and return them to Earth. Several times during its eccentric orbit about the Sun (out to about 3 AU) interstellar dust in addition to dust from Comet Wild 2 will be captured by impact into aerogel and brought back to the Earth in 2006. In addition in-situ detection and high-resolution ($M/\Delta M > 100$) compositional measurements of cometary and interstellar grains will be performed by the Cometary and Interstellar Dust Analyzer (CIDA). Interstellar dust analyses and collections may be possible even in high-Earth orbit (Grün, in preparation).

The currently studied Pluto Kuiper Express mission focuses on the big objects Pluto and Charon and perhaps one EKO, but no dust measurements in the EKB are considered. Missions to the heliospheric boundary and beyond are in their early planning phases and have to take into account dust in the outer solar system at least for hazard studies.

References

1. Eichhorn G., Grün, E. (1993) High velocity impacts of dust particles in low temperature water ice, Planet. Space Sci. 41, 429-433
2. Frisch P., Dorschner J., Greenberg M., Grün E., Landgraf M., Hoppe P., Jones A., Krätschmer W., Linde T., Morfill G. E., Reach W., Svestka J., Witt A., Zank G. (1999) Dust in the local interstellar wind, Astrophys. J. submitted
3. Grün E., Zook H. A., Fechtig H., Giese R. H. (1985) Collisional balance of the meteoritic complex, Icarus 62, 244-272
4. Grün, E., Zook, H. A., Baguhl, M., Balogh, A., Bame, S. J., Fechtig, H., Forsyth, R., Hanner, M. S., Horanyi, M., Kissel, J., Lindblad, B.-A., Linkert, D., Linkert, G., Mann, I., McDonnell, J.A.M., Morfill, G.E., Phillips, J.L., Polanskey, C., Schwehm, G., Siddique, N., Staubach, P., Svestka, J. and Taylor, A., (1993) Discovery of jovian dust streams and interstellar grains by the Ulysses spacecraft. Nature 362, 428-430
5. Grün, E., B E. Gustafson, I. Mann, M. Baguhl, G. E. Morfill, P. Staubach, A. Taylor and H. A. Zook (1994) Interstellar dust in the heliosphere, Astron. Astrophys., 286, 915-924

6. Grün E., P. Staubach, M. Baguhl, S. Dermott, H. Fechtig, B.A. Gustafson, D.P. Hamilton, M.S. Hanner, M. Horanyi, J. Kissel, B.A. Lindblad, D. Linkert, G. Linkert, I. Mann, J.A.M. McDonnell, G.E. Morfill, C. Polanskey, G. Schwehm, R. Srama and H.A. Zook. (1997) South-North and radial traverses through the zodiacal cloud, Icarus, 129, 270-288.

7. Gurnett D.A., Grün, E., Gallagher, D., Kurth, W.S., Scarf, F.L. (1983) Micron-sized particles detected near Saturn by the Voyager plasma wave instrument, Icarus 53, 236-254.

8. Gurnett D.A., Ansher, J.A., Kurth W.S., Granroth L.J. (1997) Micron-sized dust particles detected in the outer solar system by the Voyager 1 and 2 plasma wave instruments, Geophys. Res. Lett., 24, 3125-3128

9. Hanner M. S., Sparrow J. G., Weinberg J. L., Beeson D. E. (1976) Pioneer 10 observations of zodiacal light brightness near the ecliptic: Changes with heliocentric distance, In: Lecture Notes in Physics, 48: Interplanetary Dust and Zodiacal Light (H. Elsasser and H. Fechtig, Eds.) Springer-Verlag, New York, pp. 29-35.

10. Humes, D.H. (1980) Results of Pioneer 10 and 11 meteoroid experiments: Interplanetary and near-Saturn, J. Geophys. Res. 85, 5841-5852.

11. Jewitt D., Luu J., Trujillo C. (1998) Large Kuiper belt objects: The Mauna Kea 8K CCD survey, Astron. J. 115, 2125-2135

12. Krüger, H., Grün, E., Landgraf, M., Baguhl, M., Dermott, S., Fechtig, H., Gustafson, B. A., Hamilton, D. P., Hanner, M. S., Horányi, M., Kissel, J., Lindblad, B.-A., Linkert, D., Linkert, G., Mann, I., McDonnell, J. A. M., Morfill, G. E., Polanskey, C., Schwehm, G., Srama, R. and Zook, H. A. (1998) Three years of Ulysses dust data: 1993 to 1995, Planet. Space. Sci. in press

13. Landgraf, M. (1998) Modellierung der Dynamik und Interpretation der In-Situ-Messungen interstellaren Staubs in der lokalen Umgebung des Sonnensystems, PhD Thesis, University of Heidelberg

14. Liou J.C., Zook H.A., Dermott S.F. (1996) Kuiper belt dust grains as a source of interplanetary dust particles, Icarus, 124, 429-440

15. Mathis, J. S., Rumpl, W., Nordsieck, K. H. (1977) The size distribution of interstellar grains, Astrophys. J. 217, 425-433

16. Morfill G. E., Grün, E., Leinert, C. (1986) The Interaction of Solid Particles with the Interplanetary Medium, in The Sun and the Heliosphere in Three Dimensions, R. G. Mardsen, Reidel, Doordrecht, 455-474,

17. Reach, W. T. (1992) Zodiacal emission. III - Dust near the asteroid belt, Astrophys. J. 392, 289-299.

18. Stern S.A. (1996) Signatures of collisions in the Kuiper Disk, Astron. Astrophys. 310, 999-1010

19. Taylor D.A., Baggaley W.J., Steel D.I. (1996) Discovery of interstellar dust entering the Earth's atmosphere, Nature 380, 323-325

20. Witte M., Rosenbauer H., Banaskiewicz M., Fahr H. (1993) The Ulysses neutral gas experiment: Determination of the velocity and temperature of the neutral interstellar helium, Adv. Space Res. 13, (6)121-(6)130.

21. Yamamoto S., Mukai T. (1998) Dust production by impacts of interstellar dust on Edgeworth-Kuiper Belt objects, Astron. Astrophys. 329, 785-791

Simulations of Bias Effects in Kuiper Belt Surveys

Chadwick Trujillo

Institute for Astronomy
University of Hawaii
2680 Woodlawn Drive
Honolulu, HI 96822
chad@ifa.hawaii.edu

Abstract. Maximum likelihood methods are used to place limits on the allowed parameters of the Kuiper Belt. If an as yet unseen primordial Kuiper Belt exists of density 100 times higher than the known classical Kuiper Belt, it must be further than 90 AU from the Sun. Observational bias effects prevent statistically significant constraints to be placed on the size of objects such as $1996TL_{66}$, which reside in the Scattered Kuiper Belt. Bias also affects the measured maximum radial extent of the classical non-resonant Kuiper Belt.

1 Introduction

There are several theoretical reasons to think that the KBO population in the 30 – 50 AU region may be depleted compared the > 50 AU region. The gravitational influence of Neptune drops at large heliocentric distances (Levison and Duncan 1993; Malhotra 1996; Stern 1996; and Stern and Colwell 1997). In addition, collisional erosion may be reduced at these low densities (Davis and Farinella 1997; Stern and Colwell 1997; Kenyon and Luu 1998). An undepleted, primordial belt may be a factor of 100 more dense than the observed Kuiper Belt (Stern 1996). This primordial belt, if it exists, must have an inner edge lying at large heliocentric distances (> 50 AU), creating a "wall" of enhanced density in the outer Solar System. No study has constrained the parameters of this wall, based on the results of published Kuiper Belt surveys to date. This topic is addressed, and two other related questions are considered: (1) Could the Scattered Kuiper Belt objects (SKBOs) be larger, on average, than the Kuiper Belt objects (KBOs)? (2) Could there be a cutoff in the classical Kuiper Belt distribution at large heliocentric distances?

2 Maximum Likelihood Method

A model distribution of objects is created, and a "Monte Carlo" simulation "discovers" these objects in the tested survey. To be found, a body must: (1) fall on a survey field during an exposure, (2) satisfy any velocity criteria imposed in the survey, and (3) be bright enough to be detected in the survey.

Criteria (1) and (2) are tested by using the equations of apparent motion and position for an arbitrary body in orbit, as computed by Sykes and Moyni-han (1996). Criterion (3) is tested by computing an object's brightness from Jewitt and Luu (1995). Geometric albedo is assumed to be 4%.

The simulation and observations are compared by binning the distribu-tion of objects by each orbital element. Consider a single bin i, containing n_i observed objects with μ_i expected objects (as determined from the simula-tion). The probability P_i of the observations matching the drawn sample can be computed from the Poisson distribution:

$$P_i = \frac{\mu_i^{n_i}}{n_i!} e^{-\mu_i}. \tag{1}$$

The likelihood L_a, considering for example, the single orbital element a, that one realization drawn from the model will match the observations is the joint probability of P_i for all i.

$$L_a = \prod_{i=1}^{n} P_i \tag{2}$$

This procedure is repeated for each orbital element, and the product is com-puted to give the total likelihood. Since this value is very small (typically of order 10^{-100}), the natural log of the probability is calculated in order to eliminate the possibility of floating point precision errors. Bin sizes are chosen such that binning effects are negligible.

3   The Primordial Kuiper Belt

Stern (1996) suggests that the primordial Kuiper Belt, if it exists, may have 100 times the density of the known Kuiper Belt. This model is adopted here. The widest KBO survey published to date (Jewitt, Luu, and Trujillo, 1998, hereafter JLT 1998) was tested, with the addition of another 25 sq deg, for a total of 75 sq deg surveyed to limiting red magnitude $m_R = 22.5$. A fit was done to the discovered classical belt KBOs, chosen as those KBOs observed on more than one opposition, with semimajor axes obeying $43 < a < 45$, avoiding the major Neptune resonances (Malhotra, 1996). The best fit orbital elements from this procedure were used to create the simulated distribution, but the surface density was increased by a factor of 100, and semimajor axes were increased according to Table 1, to simulate a massive, distant belt. The size distribution was assumed to follow a differential power law,

$$n(r)dr = \Gamma r^q dr, \tag{3}$$

where $n(r)dr$ is the number of objects with radii between r and $r + dr$, and Γ determines the density of objects.

Results are shown in Figure 1. For a q=-4 size distribution (the best fit of JLT 1998) truncated at small radii, such as the $r < 125$ km case, there

Table 1. Model parameters. $a, e, i, \omega, \Omega, M$ represent the Keplerian orbital elements: semimajor axis, eccentricity, inclination, argument of perihelion, longitude of the ascending node and mean anomaly, respectively. $\Sigma(R)$ is the ecliptic plane surface density as a function of heliocentric distance.

element	min	max	distribution
a	60–100 AU	250 AU	$\log(\Sigma(R)) \sim -2$
e	0	0, 0.2	uniform
i	0	$30°$	uniform
ω, Ω, and M	$-180°$	$180°$	uniform
radius	100 km	125, 250, 500, 1000 km	$q = -4, -5$
N	10^6	10^6	N/A

is no constraint, because even at perihelion, these objects are too faint to be detected. The most likely case, with maximum radius about equal to Pluto's size ($r < 1000$ km) requires $a_{min} > 90$ AU, with little dependence on the maximum eccentricity. Tests were also done for q=-5 (the best fit for Gladman, et al.), resulting in $a_{min} > 85$ AU for $r < 1000$ km. Thus, with reasonable choices of size distribution and maximum eccentricity, the inner edge of the primordial Kuiper Belt must lie beyond about 90 AU, assuming that the density contrast is a factor of 100.

A simulation was run to determine the sky coverage necessary for each of several survey types to detect a primordial Kuiper Belt beginning at 90 AU and extending outward. For this computation, the assumed size distribution followed q=4. In the literature, two types of surveys dominate: wide-field surveys and very deep surveys. Three hypothetical models were considered, representing the extrema of surveys performed: (1) the wide-field survey previously discussed (JLT 1998), with limiting red magnitude $m_R = 22.5$; (2) The Keck "medium-deep" survey (Luu and Jewitt 1998, hereafter LJ 1998) with $m_R = 26.1$; and (3) The Keck "ultra-deep" survey (LJ 1998) with $m_R = 26.6$. The results appear in Table 2.

As expected, the deepest survey requires the smallest amount of sky area to detect the primordial KBOs, should they exist. However, the wide-field survey is the most efficient; although the wide-field and medium-deep surveys both require about the same amount of total integration time T, the medium-deep survey requires a 10-meter telescope, while the wide-field only requires a 2.2-meter telescope.

4   Scattered Kuiper Belt Objects

One basic observation of the SKBOs is that the only member known (1996TL$_{66}$) has a radius of about 200 km (Luu et al., 1996), which is twice that of the typical KBO (Jewitt, Luu, and Chen 1996). This suggests that the SKBOs are substantially larger than the KBOs.

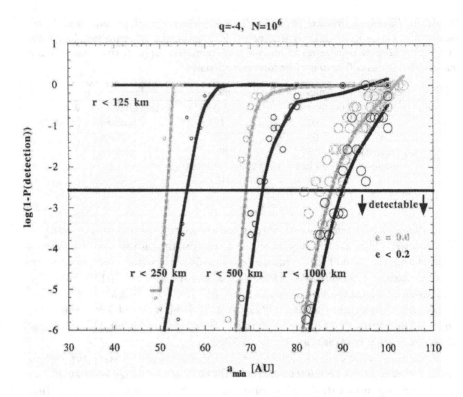

Fig. 1. The likelihood of non-detection versus minimum semimajor axis of the simulated primordial belt distribution. The maximum semimajor axis was fixed at 250 AU. Points lying below the horizontal line represent models that are ruled out at the 3σ level. Data point scatter is due to different realizations of the models, and the solid lines represent the data after smoothing.

Table 2. A comparison of hypothetical surveys sensitive to a belt of enhanced KBO density beyond 90 AU. A represents the amount of sky coverage necessary to place the belt in the 3σ detectable regime. D, t, n, and T represent telescope diameter, exposure time for a single image, number of images to cover A, and the total integration time of the survey, respectively.

survey type	m_R	A [sq deg]	D [m]	t [s]	n	T [hr]	ref
wide-field	22.5	100	2.2	150	1100 × 3	140	JLT 1998
medium-deep	26.1	2	10	900	170 × 3	130	LJ 1998
ultra-deep	26.6	1	10	4500	90 × 3	340	LJ 1998

To test the statistical significance of this possibility, a distribution of SKBOs was created, with a, e, and i matching 1996TL$_{66}$, but with ω, Ω, and M unconstrained. A q=-4 size distribution is assumed, with 100 km $< r < 1000$ km. A histogram of discovered SKBO size is plotted, as shown in Figure 2. The mode of this distribution, corresponding to the most probable SKBO size, lies at $r \sim 200$ km, regardless of binning method. Thus, the large observed SKBO size is not significant.

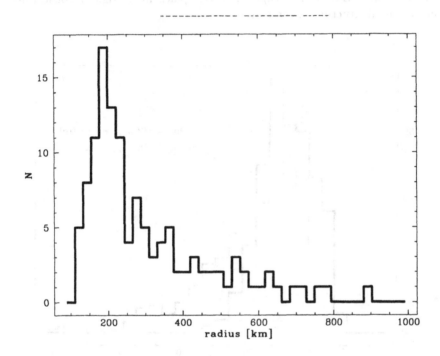

Fig. 2. The simulated size of SKBOs at discovery, assuming a q=-4 size distribution.

5   Truncated Kuiper Belt

Another observation of interest is that wide field surveys ($m_R < 22.5$) have found KBOs at about the same heliocentric distances as very deep surveys ($m_R < 26.6$) which are over 4 magnitudes more sensitive. This suggests that there may be an outer truncated "edge" to the classical Kuiper Belt beyond about 50 AU, the furthest discovery distance.

To probe the statistical significance of this theory, the wide field survey of JLT 1998 was compared to the deepest Keck survey of LJ 1998. For each, a histogram of discovery distance was plotted assuming the existence of a classical belt extending beyond 100 AU, as seen in Figure 3. The probability that the deepest survey would show no greater discovery distance than the wide field survey is determined. With 14 KBOs discovered in the wide-field survey and 1 KBO discovered in the deepest Keck survey this probability is about 70%, which is ~ 1σ. In this limited sense, and within the assumptions of the model, I find no statistical basis for thinking that the classical Kuiper Belt is truncated. New University of Hawaii observations underway at the Canada-France-Hawaii Telescope, however, place much tighter constraints on the radial structure of the belt.

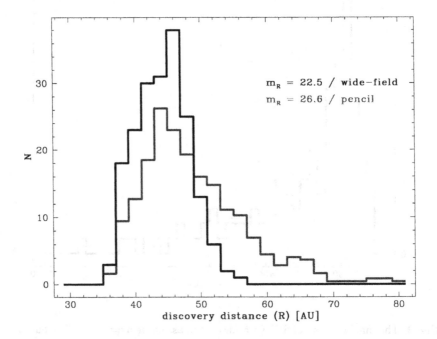

Fig. 3. The simulated discovery distance of two types of surveys.

6 Summary

Monte Carlo simulations and maximum-likelihood methods are used to probe the bias effects of KBO surveys. The following results are presented:

(1) If there exists a massive primordial belt of approximately 100 × the density of the current Kuiper Belt, it lies beyond 90 AU.

(2) The most efficient survey method to probe the existence of the hypothetical massive primordial belt is a wide-field survey.

(3) The only known SKBO ($1996TL_{66}$) is about a factor of 2 larger than the typical KBO. However, this is to be expected due to survey bias effects.

(4) The deepest ground-based survey to date and the widest-field survey differ by 4 magnitudes in sensitivity, however both discover objects at about the same heliocentric distance. This apparent inconsistency is not formally indicative of an edge to the Kuiper Belt at 50 AU.

References

1. Davis, D. R. and Farinella, P. (1997): Icarus, 125, 50.
2. Jewitt, D. C. and Luu, J. X. (1995): AJ, 109, 1867.
3. Jewitt, D. C. and Luu, J. X. (1998): ApJ, 502, L91.
4. Jewitt, D., Luu, J., and Chen, J. (1996): AJ, 112, 1225.
5. Jewitt, D. C., Luu, J. X., and Trujillo, C. (1998): AJ, 115, 2125.
6. Kenyon, S. J. and Luu, J. X. (1998): AJ, 115, 2136.
7. Malhotra, R. (1996): AJ, 111, 504.
8. Levison, H. F. and Duncan, M. J. (1993): ApJ, 406, L35.
9. Luu, J., Marsden, B. G., Jewitt, D., Trujillo, C. A., Hergenrother, C. W., Chen, J., Offutt, W. B. (1997): Nature, 387, 573.
10. Stern, S. A. (1996): AJ, 112, 1203.
11. Stern, S. A. and Colwell, J. E. (1997): ApJ, 490, 879.
12. Sykes and Moynihan. (1996): Icarus, 124, 399.

A Pencil-Beam Search for Distant TNOs at the ESO NTT

Hermann Boehnhardt[1], Olivier Hainaut[1], Catherine Delahodde[1], Richard West[2], Karen Meech[3], Brian Marsden[4]

[1] European Southern Observatory, Alonso de Cordova 3107, Santiago de Chile
[2] European Southern Observatory, Karl-Schwarzschild-Str. 2, D-85748 Garching, Germany
[3] Institute for Astrophysics, 2680 Woodlawn Drive, Honolulu, HI 96822-1839, USA
[4] Harvard-Smithsonian Center for Astrophysics, 60 Garden Street, Cambridge, MA 02138, USA

Abstract. We used the image data set obtained during our NTT/EMMI observing campaign of the TNO 1996 TO_{66} at ESO La Silla (see our paper presented during this workshop) to start a pencil-beam search for very distant TNOs. During our 5 observing nights the exposures of the foreground TNO were centred at the same target position on the sky (apart from a small amplitude jittering). For the sampling of the lightcurve of 1996 TO_{66} a long series of R filter exposures was taken during 4 of the 5 nights. The individual R filter exposures of 15 min each reach a limiting brightness of about 24.5 mag, the aligned (using the background stars) and coadded R images of a single night go down to 26 mag. By blinking individual and coadded images a pencil-beam segment of 8.5 × 8.5 arcmin could be searched for unknown solar system objects (several main belt asteroids were easily recognized). The blinking of the individual frames should have allowed us to identify TNOs with a typical diameter of 100 km at about 50 AU solar distance; the same procedure applied to the co-added images should allow to find objects of the same size at about 80 AU or 400km bodies at 150 AU distance.

This simple pencil-beam search in our narrow EMMI field of view was not successful, i.e. no distant TNO was found. According to Jewitt, Luu and Chen's model assumption (Astron. J., 112, 1225-1238, 1996), one would expect about 0.2 TNOs in such a field, so our negative detection is not a surprise. However, refined search techniques are under development and, furthermore, the use of the 0.5 deg Wide Field Imager at the 2.2m telescope in La Silla will greatly improve the statistical significance of such deep search programmes.

1 A Search-Included TNO Programme

In late October 1997, a 5 night observing programme at La Silla Observatory targeted the Transneptunian object (TNO) 1996 TO_{66}. The main goal of these observations was the physical characterization of this distant solar system body by means of multi-colour photometry and low-dispersion spectroscopy (for the results see the 'A Portrait of 1996 TO_{66}' by Delahodde et al. presented at this workshop). However, it also contained a small pencil-beam search for unknown and more distant solar system objects utilizing the series of CCD images taken for the photometric rotation period analysis of 1996 TO$_{66}$. In order to accomplish both goals with the same data set, i.e. the photometry of the rotational variability and the pencil-beam search for new objects, we tailored the concept for the observations such that we pointed the telescope to the same area of the sky while the TNO was moving very slowly (order of a few arcsec/h) across this region during a period of several nights. The number of nights during which the target TNO can be imaged on the same star background is basically determined by the field of view of the telescope/instrument equipment and by the velocity of the object used for the photometry. Of course, a 2-3m class telescope is needed to achieve accurate magnitudes for the 21-22 mag target and an even fainter limiting magnitude for the search programme. Although the detection of several foreground moving objects from the Asteroid Main Belt could be expected, the main aim of the search programme was to look for more distant solar system objects, and in particular for TNOs beyond 50 AU from the Sun. In the following sections we describe in detail the observing strategy, the reduction technique applied to the data and the results of the search (at least of our first attempt). We finally discuss the requirements for our planned future search projects for objects in the outer solar system.

2 The Observation Strategy

The observations were performed during 21-24 Oct. 1997 at the 3.6-m New Technology Telescope (NTT) at the European Southern Observatory in La Silla/Chile (the fifth night - 25 Oct. 1997 - scheduled for this programme was used for spectroscopy of the foreground TNO 1996 TO$_{66}$ and did not contribute to the image set of our search programme) using the EMMI multi-mode instrument at the Nasmyth focus of the telescope. The 'red' arm of the instrument allowed us to obtain BVR filter CCD imaging over a 9.2 × 9.2 arcmin field of view (the Tektronix CCD has 2048×2048 pixels of 24 μm size which corresponds to 0.27 arcsec on the sky). For the sampling of[°] the rotation period of 1996 TO$_{66}$ and for the collection of a large number of images for the deep search project, mostly broadband Bessel R filter images were taken (a few Bessel B and V filter images were obtained for the determination of the colour index of 1996 TO$_{66}$). The typical exposure time of the exposures was 15 min. The sky was photometric during most of the observing run with typical seeing better than 1 arcsec, and the moon was below the horizon during the TNO observing window for La Silla.

The pencil-beam search project required the following observing strategy:

1. to observe the same target field through the same filter with a very long total exposure time in order to increase the limiting magnitude for the

search (also good seeing conditions are helpful for the success of the search)

2. to use sidereal tracking for the imaging in order to allow the proper alignment of the images during the data analysis (no tracking [°]of the foreground TNO because of the refraction effects in the star images)

3. to jitter the telescope pointing in the target area for the superflating of the images and the removal of fixed patterns of the CCD detector

Long total integration times in a short observing period (as typically the case for many ESO programmes) can best be achieved if the object is in or close to opposition to the Sun when the observing window per night is longest. Apart from that, in our cas[°]e the constraining factors for requirement 1 were the observing programme committee (OPC) of ESO (which decides on the observing time of the ESO proposals) and the motion of the foreground TNO for which we wanted to collect CCD photometry in the same field of view of the instrument. At the time of our observations the prime target 1996 TO_{66} moved with about +2.3 arcsec/h in right ascension and about -0.8 arcsec/h in declination. The typical transit time of this TNO for the EMMI field of view was about 10 days and, thus, the ESO OPC was somehow the limiting factor for our search project, since it allowed us only 4 half nights of observations, i.e. we had an observing window of about 3.25 days total duration.

Although the EMMI field of view covers roughly the 3.25 days track length of a solar system object at about 17 AU distance from the Sun, it is obvious that the success of the search for unknown objects in the instrument field of view by coadding the data from several nights (as described in section 4) will suffer from the loss of objects when they are located close enough to the edge of the image such that they can exit or enter the field of view of the instrument during the observing window. Thus, the 'horizon of complete coverage' of an instrument can be defined as the minimum distance of an object in a circular orbit around the Sun that stays in the instrument field of view for a given time interval if it is found there at the beginning of the observations. For our 3.25 days EMMI observations the 'horizon of complete coverage' was nominally about 40 AU (which is slightly less than the distance of our prime target 1996 TO_{66} at 45.7 AU from the Sun at the time of our observations), but effectively it was a few AU more because of the shrinking of the viewing field (due to the jittering mode used for our observations; see requirement 3 and discussion below).

For the search of distant and slowly moving TNOs it is advantageous to use sidereal tracking since a priori the velocity and direction of the potentially detectable objects are unknown and the trailing effects will be small during a 15 min exposure (for our prime target 1996 TO_{66} it was 0.6 arcsec, which was smaller than or of the order of the typical seeing during our observing interval). Beyond that the influence of the atmospheric refraction on the direction and length of star trails can be avoided. This is important if one aims for a removal of background objects using methods as described in Boehnhardt et al. (1997) and Hainaut et al. (1998).

Requirement (3) in principle widens the useful area for the search, but at the same time it reduces also the detection limit in the outer regions of the target field, since due to the jitter amplitude they are not longer covered by the same number of exposures (or in other words they differ in their total exposure time). For our imaging we applied a telescope jittering of 20 arcsec, i.e. in between the individual exposures the telescope pointing was offset in both directions according to a random-walk sequence with a maximum amplitude of 30 arcsec from the field centre. Thus, the total search area increases to about 9.5×9.5 arcmin, but the area of optimum signal-to-noise (S/N) for the search decreases to about 8.9×8.9 arcmin.

The coordinates of the target field for our search were close to the mid-time position of the prime target for our photometric part of the observing programme, i.e.

Right Ascension (2000) 23h 52m 32s
Declination (2000)   +01° 42' 30"

The observations were executed as a series of 15min R filter exposures per half night, interrupted only by a few images through the B and V filters (for the colour determination of 1996 TO_{66}). During the 4 nights, 36 R filter images of the field were taken which gives a total integration time of 9 hours. Following requirements (2) and (3) telescope jittering and sidereal tracking were exercised for this programme. For the data reduction and flux calibration photometric standard stars (Landolt 1992), as well as bias, dark and twilight flatfield images, were obtained during the observing run.

3   The Data Reduction

The basic data reduction comprised bias and dark current subtraction, flatfield division, cosmic ray removal and photometric calibration. During the bias correction a 2dim bias frame was subtracted that was derived from a master bias constructed from the series of available bias exposures of this run and scaled to the bias level of the respective flatfield or object images using the overscan region of the frames. The flatfielding procedure followed the recipe of Hainaut et al. (1998), which uses both the object frames and twilight flats to calculate the optimum flatfield. This procedure delivers relatively calibrated images with background variations across the field of view of below 1 percent for large scale structures and below 0.1 percent for small scale variations. For the cosmic ray removal a standard routine of the ESO MIDAS image processing package was applied. The photometric calibration of the object frames was achieved using the standard star exposures. First order extinction and colour corrections were applied during the photometric reduction. Because of the excellent field overlap of the TNO exposure series all object frames could be flux calibrated (using relative photometry of field stars in case of images that suffered from rare, slightly non-photometric sky conditions during the observing run).

In a second step the object images were aligned to sub-pixel accuracy, i.e. the field stars had the same xy pixel coordinates in all images to within sub-pixel accuracy. Thereafter, the R filter images of a single night were coadded in stacks of 4-5 subsequent exposures. In that way, an increase of the S/N ratio of faint and unknown solar system objects in the field of view can be expected if their motion rate is not too fast that their images are smeared out along a longer trail length.

4   The Results with a Short Discussion

The search for moving objects in the frame was exercised through blinking two or more single and coadded images of the target field on the computer. Moving objects could thus be identified visually by the change in the pixel position in between the blinked exposures. Different combinations of images from the sets of single and coadded frames were used for the blinking. However, apart from several objects in the Main Asteroid Belt (which were already recognized as trailed objects in the images during the observing run) no further moving object was detected either in the single or in the coadded images.

The limiting magnitude of about 24.5 mag for single frames translates to an object radius of about 100km at 50 AU solar distance (assuming an albedo of 0.04), the 26 mag limit for the coadded frames to the same object radius at 80 AU or to a radius of 400 km at 150 AU. Since our observations fulfill the criterion for the 'horizon of complete coverage' for objects beyond about 40 AU from the Sun, we conclude that there were no slowly moving objects in the outer solar system in our EMMI field of view at the time of observations. This applies to objects with suitable size for detection with EMMI (i.e. to be brighter than 24.5 mag) with an upper limit for extremely distant bodies which are not longer recognizable by blinking, since they are too far away and thus do not move more than about 1 arcsec during our whole observing interval of 3.25 days (i.e. objects beyond about 10000 AU). A similar conclusion for the 26 mag limit of the coadded frames can only be done with caution since, from the point of view of data reduction, we may not have achieved the optimum S/N for moving objects due to our simple coaddition of frames which are aligned for the stars (see discussion in section 5).

Applying Jewitt et al.'s (1996) assumption about the object distribution and sizes of TNOs to our observing conditions (EMMI field of view, limiting magnitude of 26 mag in R), we arrive at a probability of 20 percent for the detection of an Edgeworth-Kuiper-Belt (EKB) object in our data. Therefore, the non-detection we have as a result of our data analysis (although biased by the fact that the coadded images are not corrected for anticipated motion of the objects searched for) is not too much a surprise. Unfortunately, it does not provide much constraint on the structural and size models of the EKB.

5 Our Plans for Doing Better

The search for TNOs in our EMMI data as described above was only a by-product that we could get for free from our imaging campaign of the prime target 1996 TO$_{66}$. Although the outcome of the TNO search was negative, we find it a challenging enterprise to do better - and more work in this area of research - in the future. So, what are our plans?

Distance range scanning by shift-and-add examination of the images: of course, the coaddition of frames aligned to the star background is too simple to achieve the optimum S/N for the detection of slowly moving objects in our EMMI images of 1996 TO$_{66}$. The images should be systematically shifted in x and y assuming a particular speed and direction of motion for the objects being sought. Thereafter, the images should be coadded to improve the S/N of the potential TNO, the image of which should then fall in the same pixel area in all frames. The object detection can be done by looking for a point source in the data while the stars will appear as trails. However, blinking two or more such images may help to ascertain the identification. The velocity parameter for the shift basically represents a certain range of heliocentric distance for the sought objects. One can thus systematically scan through the whole outer solar system looking for new distant objects in the same data set. At present our EMMI data of 1996 TO$_{66}$ are evaluated by applying such a shift-and-add search algorithm (colaboration with Watanabe group in Japan).

Background object removal: the shift-and-add approach, as well as the use of deep exposures, leads to a crowding of the images by background objects (either stars, or more normally galaxies, are the objects of concern). This implies that the useful pixel area (i.e. free from background objects) gets smaller, the more images are coadded and/or the deeper the exposures are. Therefore, we are thinking of the removal of background objects in a more self-consistent way, i.e. by using the image data themselves. Such processing for the removal of background objects has successfully been applied in the past for the detection of very faint comets on a crowded background (Boehnhardt et al. 1997, Hainaut et al. 1998), and one can expect that it will work for the distant TNO search also.

Wide and deep field searches: at present new telescope and CCD technology is being introduced world-wide that will hopefully advance the exploration of the outer solar system in a quantum step (as is promised also for other fields of astronomy). Wide-field CCD imaging of 0.5 to 1 deg is becoming available at 2-3m class telescopes at different observatories (ESO, Hawaii, La Palma). These telescopes will certainly allow one to reach 24-26 mag limiting magnitude in wide area search campaigns for TNOs that will fill in the gaps that we have right now in the wide-field coverage of the Kuiper-Belt region. Beyond 25-26 mag, pencil-beam searches on 8-10m class telescopes (FORS at the ESO VLT, LRIS at Keck, FOCAS at Subaru) can explore more in a "spotlight" approach the greater distances and a smaller range of the solar system bodies, down to 30 mag and beyond. The ultimate step, however, will be achieved if we can go wide and deep at the same time, i.e. if wide-field CCD imaging becomes possible on these large telescopes. VMOS at the ESO VLT and the prime focus camera of the Subaru telescope will be the workhorses for such applications.

As a final remark, we should like to point out that all approaches mentioned above will have one point in common, i.e. they will need significant computing power and disk storage for the data reduction and evaluation. An image of the 0.5 deg Wide Field Imager (WFI) at the 2.2m telescope at ESO La Silla (this instrument will become operational in January 1999) is about 130 MegaBytes of FITS data, and a single observing night at the WFI is expected to deliver about 20-30 GigaByte of FITS images. Numerical manipulation of such images requires fast computers with large RAM storage to deal with the images efficiently (i.e. in a finite time of a few hours or so). The ultimate step in this respect will be the automatic detection of moving objects in wide-field imaging data using computers and sophisticated search algorithms on TeraBytes of data.

References

1. Boehnhardt, H., Babion, J., West, R.M., 1997, A&A 320, 642
2. Hainaut, O.R., Meech, K.J., Boehnhardt, H., West, R.M., A&A 333, 746
3. Jewitt, D.C., Luu, J.X., Chen, J., AJ 112, 1225
4. Landolt, A.U., AJ 104, 340

Does Pluto Affect the Trans-Neptunian Region?

Brett Gladman[1], Jean-Marc Petit[1] and Martin Duncan[2]

[1] Observatoire de la Côte d'Azur, Departement Cassini, B.P. 4229, 06304 Nice Cedex 4, France
[2] Queen's University, Dept. of Physics/Astronomy, Kingston On, K7L 3N6, Canada

Abstract. We present a short analytic and numerical calculation showing that Pluto's gravitational effect produces negligible dynamical excitation over the age of the solar system.

1 Introduction

During the MBOSS 98 meeting D. Jewitt raised the question of whether Pluto's gravitational influence could have stirred up eccentricities and inclinations of trans-neptunian objects (TNOs) in (1) the classical disk from 40–50 AU, (2) the Neptune trojan region, and (3) the 2:3 resonance region. The idea is that although Pluto's mass is small (0.003 Earth masses, escape velocity of v_{esc}=1.3 km/s), it loops through the belt for 4.5 Gyr and perhaps its cumulative effect is noticable. Several MBOSS participants claimed that this was clearly not the case, and could be proved with a short analytic calculation. We present this result and numerical evidence to support it.

2 The Back-of-the-Envelope Calculation

Pluto can produce a maximum velocity change of order its escape velocity only on objects skimming its surface (actually, the maximum scattering gives 70% of v_{esc}). A 1 km/s perturbation on a circular orbit at 40 AU (where orbital speeds are 5 km/s) could produce an eccentricity of 0.2, about the right magnitude needed. Let us be optimistic, and determine the fraction of the belt that Pluto has passed within 3 planetary radii (which would give maximum velocity kicks of $\simeq$500 m/s, or $e \sim 0.1$). This fraction is just the volume V_P swept out by a disk of 3 Pluto-radii moving at its velocity *relative* to the belt over the age of the solar system, divided by the volume between its perihelion and aphelion at 30 and 50 AU. There are two ways of expressing V_P. The first is:

$$V_P = \pi(3R_{Pluto})^2 \times v_{rel} \times Age_{ss} \tag{1}$$
$$\sim \pi \, 10(10^3 km)^2 \times 1 \, km/s \times 1.4 \times 10^{17} sec \sim 5 \times 10^{24} \, km^3 \,, \tag{2}$$

where v_{rel} is Pluto's velocity relative to the belt, of order $ev_{orb} \simeq 0.2(5km/s)$ $= 1km/s$. This same result can be obtained by multiplying the volume traversed by a single orbit of Pluto by the number of orbits Pluto has made in 4.5 Gyr. The volume between 30 and 50 AU within 20 degrees of the ecliptic is roughly $5 \times 10^{29}km^3$. Thus, Pluto gives $e \sim 0.1$ to only 1 in 10^5 of the TNOs in the classical region, over the lifetime of the solar system.

This is over-optimistic for the objects now known (*i.e.*, near the ecliptic), since Pluto spends very little time there due to its 17-degree orbital inclination. The situation is even worse for the Neptune Trojans, since Pluto is locked in the Kozai resonance, amongst others [1], which prevents its perihelion from ever being near the ecliptic, and thus it never comes within several AU of the core of a hypothetical Neptune-Trojan cloud.

The situation for TNOs in the 2:3 resonance with Pluto is less clear. To kick TNOs out of the resonance, Pluto must produce a $\Delta a/a \sim$ the resonance width ~ 0.2 AU, so $\Delta a/a \sim 5 \times 10^{-3}$. Therefore, the distance of minimum passage from Pluto grows from 3 to about $3 \times (0.1/0.005) = 60$ planetary radii, increasing V_P by a factor of 400. But, V_P drops since v_{rel} is smaller in the volume of the 'target region'. Thus V_P increases by something like 2 orders of magnitude. Can the 'target volume' be decreased? Although 2:3 objects have similar semimajor axes and roughly similar e's, this will not reduce the target volume by the required additional factor of 10^3, especially considering the rather large range in inclinations and libration amplitudes. Thus, it is unlikely that Pluto can perturb any significant fraction of the 'plutinos' out of the resonance over the lifetime of the solar system.

3   Numerical Calculation

To potentially improve on the above calculation, the orbit of Pluto was integrated and recorded for 10 Myr under the gravitational interaction with Neptune (only); this should be long enough to capture the important long-term variations of its orbital elements. This orbital history was then fed through the close-encounter code described in Petit *et al.* [2], which calculates the excitation that will be produced on 10%, 50%, and 90% of the test particles in a disk of 1-degree vertical extent in the region of interest (here, 30 to 50 AU).

The results were then simply scaled to the age of the solar system. As Figure 1 shows, if Pluto alone was responsible for the excitation of the classical belt, the median excitation would be 10^{-4}. Here the 'excitation' indicates the order of the final eccentricities and inclinations of the TNOs, or the fractional semi-major axis changes that could have occured. It is clearly negligible. The numerical calculation also confirms that Pluto can do nothing to the Neptune Trojan region, since it never comes within 1 degree of the ecliptic near 30 AU.

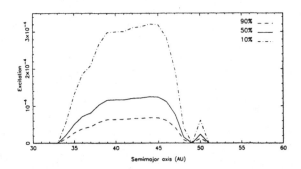

Fig. 1. The 'excitation' produced by Pluto on a 1-degree vertical disk between 30 and 50 AU. The ordinate is the magnitude of the $\Delta e \sim \Delta i \sim (\Delta a/a)$ felt by 10%, 50%, and 90% of the TNOs in that disk.

From limited direct numerical integrations of the 4 giant planets, Pluto, and particles placed in the 2:3 resonance, we confirm that Pluto has no detectable effect at emptying the 2:3 resonance.

(Yu, Q. and Tremaine, S. (1999) Astron. J., 118, 1873-1881 have independently examined the effect of Pluto on the dynamics of the Plutinos. They find that Plutinos with fractional eccentricity differences $0.1 < \Delta e/e_p < 0.3$, where $e_p = 0.25$ is the eccentricty of Pluto, are liable to become Neptune-crossing as a result of perturbations from Pluto and are therefore dynamically unstable - Eds.)

References

1. Milani, A., Nobili, A., Carpino, M. (1989) The Dynamics of Pluto. Icarus, **82** 200–217.
2. Petit J-M., Morbidelli, A., Valsecchi, G. (1998) Large scattered planetesimals and the excitation of the small body belts. Icarus, **141** 367–387.

Kiso EKBO & Centaur Survey and the Design and Implementation of the Moving Object Detection Engine

Daisuke Kinoshita[1], Naotaka Yamamoto[1], Tomohiko Sekiguchi[2], Shinsuke Abe[2], and Junichi Watanabe[3]

[1] Science University of Tokyo, Shinjuku, Tokyo, 162-8601, Japan
[2] Graduate University for Advanced Studies, Mitaka, Tokyo, 181-8588, Japan
[3] National Astronomical Observatory of Japan, Mitaka, Tokyo, 181-8588, Japan

Abstract. We are now carrying out a survey of Edgeworth-Kuiper Belt Objects (EKBOs) and Centaurs using the 1.05-m Schmidt telescope with a 2K CCD camera (2048 × 2048 pixels) at the Kiso Observatory, Japan. We call this project the "Kiso EKBO & Centaur Survey". The field of view is 48' × 48' and the R-band limiting magnitude is $m_R = 21$ for this system. The aim of this survey is to determine the bright end of the luminosity distribution of EKBOs which provides the time scale of planetesimal growth and to detect EKBOs and Centaurs that are bright enough for physical observations. Due to its large field of view, we have already surveyed more than 2.6 square degrees and still continue this project. In order to search for objects in the images, a detection program for distant minor bodies was developed. We report an overview of Kiso EKBO & Centaur Survey along with the design and implementation of our "Moving Object Detection Engine" including its application to our data.

1 Introduction

Surveys in this decade have revealed the existence of a large number of minor bodies in the outer solar system. Since 1992, more than 70 Edgeworth Kuiper Belt Objects (hereafter, EKBOs) have been found in the region beyond Neptune. These are the objects that Edgeworth (Edgeworth [1]) and Kuiper (Kuiper [3]) had predicted.

The cumulative luminosity function between 20 and 26.6 magnitude is well fitted by a power law distribution(Luu and Jewitt [4]). However, the bright-end of the luminosity function is still uncertain. It is possible to determine the time scale of planetesimal growth if the bright-end of the luminosity function of EKBOs is known (Jewitt et al. [2]). It is also important to discover bright EKBOs to carry out physical observations.

2 Kiso EKBO & Centaur Survey

The Kiso EKBO & Centaur Survey started in April 1998, using the 1.05-m Schmidt Telescope with a 2K x 2K CCD camera at the Kiso Observatory

(University of Tokyo). Our survey is suitable to obtain the luminosity function around $R \sim 21$ mag. The 1.05-m Schmidt Telescope has a large field of view, 48 arcmin × 48 arcmin, or 0.64 sq. degrees.

We attempted to perform our survey twice in 1998. From 17 to 21 April, we searched 2.6 sq. deg. but from 17 to 20 August, we could not take any data for the survey due to both software and hardware trouble at the observatory. To date, we have not found any EKBOs or Centaurs. However, the expected number of EKBOs in our survey area is only 0.2 according to the cumulative luminosity function obtained by Luu and Jewitt [4]. We plan to continue the Kiso EKBO & Centaur Survey in the future.

3   Automated Detection Software

We have developed automated detection software for moving objects in digital images called Moving Object Detection Engine (MODE). The design of our detection program is similar to the software called MODS (Trujillo and Jewitt [5]). First it lists all of the stars in three images separated in time, then it matches pairs of field stars based on their relative positions. Objects that are not paired are candidates for EKBOs or Centaurs. It checks if the candidates are EKBOs by calculating their direction, motion, and flux. MODE has the advantage of using a simple and fast argorithm called "SIFTER". When we assume the heliocentric distance of the objects, the SIFTER shifts the images and takes the image correlation to find the moving objects. This procedure is possible because the apparent motion of distant minor body is decided primarily by the motion of the Earth. MODE is implemented on the "FreeBSD" operating system that runs on a PC. This software was applied to the images obtained by our survey. We have succeeded in detecting known main belt asteroids automatically. Currently, we are improving the performance of MODE with features, such as, accurate sky background estimation, detection of low signal-to-noise ratio objects, and precise correction of pointing error.

References

1. Egeworth, K., 1949, MNRAS, **109**, 600.
2. Jewitt, D., Luu, J., Trujillo, C., 1998, AJ, **115**, 2125.
3. Kuiper, G. P., 1951, in Astrophysics, ed. J. A. Hynek (McGraw-Hill, New York), p. 357.
4. Luu, J., Jewitt, D., 1998, ApJ, **502**, L91.
5. Trujillo, C., and Jewitt, D., 1998, AJ, **115**, 1680.

Limitations of Numerical Modelling of Kuiper Belt Object Dynamics

Ryszard Gabryszewski

Space Research Centre, ul. Bartycka 18 A, PL 00 - 716 Warszawa, Poland.
email: r.gabryszewski@cbk.waw.pl

Abstract. The investigation of the dynamics of KBOs is based on numerical orbital integrations on extremely long time scales due to orbital evolution of particles. The evolution of KBOs to JFCs needs a time-span of the order of 10^9 years. Such a long time of integration affects the errors. So the question arises, what is the boundary of an integration time to distinguish the physical solution from numerical noise and what it depends on. This paper presents numerical integrations of 150 massless test particles in a model of the Solar System with 4 giant planets and the central mass. For each test particle computations were repeated at least twice on different computers. The results show that an increase of errors in a solution depends on the eccentricity and the inclination of an orbit. The estimated maximum time-span of integration is of the order of 10 million years for highly elliptical orbits ($e \sim 0.6$) and up to 125 million years for quasi-circular orbits (for a particular model of the Solar System with orbits of massless objects outside Neptune's orbit). After a long time-span of integration (120 - 130 Myrs), the solution can be completely chaotic. It cannot be stated unequivocally that this is one of the possible paths of particles or that this is just numerical noise. So a different way of studying dynamical evolution of KBOs and SP comets is needed. The integration of the equations of motion between particular phases of objects which are considered as comets in different phases of their lives (KBOs - Centaurs - Comets - possibly extinct Comets) could be a new way of studying the dynamical evolution of SP comets.

1   Introduction

In a process of numerical integration, errors always affect the final solution. For n-body problem integrations over long time scales, there is a possibility that numerical errors dominate the natural solution. So one of the most important question concerns the maximum time interval (the maximum time-span) of integration, after which the obtained solution is scientific, natural, and not just a numerical fluctuation. This is one of "boundary conditions" to be defined before making any numerical modelling. Three types of error can be named, which are able to strongly affect the precision in an integration process:

- an error connected to the method of integration,
- an error connected to the way of doing arithmetic operations by a processor, and

- an error connected to a kind of property of equations of motion: deterministic chaos.

These errors can only be diminished but there is no way to get rid of them. Accumulation of errors in a numerical process can cause that solution to only be a numerical fluctuation after a certain time-span. It may be assumed that one of the simplest tests to distinguish between noise and a natural solution could be a test for repetivity of the results. The maximum interval time can be found by integrating the same n-body problem with the same initial conditions on two or more different computers having the same bit - representation of numbers. This maximum time span can be estimated as a period during which the investigated evolutions are the same or similar.

2   Calculations

150 test particles generated from orbits of two different KB objects, 1992 QB_1 and 1996 TL_{66}, were tested. The generation of test particles was performed as follows: a solar system model was assumed as a central mass and 4 giant planets. Equations of motion of 150 massless test particles were integrated over 200 to 500 Myrs. Orbits of the particles were generated as follows:

- 1992 QB_1 and 1996 TL_{66} Kuiper Belt objects were chosen,
- orbital parameters of these objects were calculated using all available observations,
- orbital parameters were copied 80 times for $1992QB_1$ and 70 times for $1996TL_{66}$), and then
- modified to obtain new, orbital parameters (gaussian distribution), and
- all these orbits were integrated on two different machines using:
  - the Bulirsch - Stoer method, and
  - the RMVS3 method - the (RA15 method).

The initial conditions for the planets are based on DE118. The integration of equations of motion were performed using three methods: first two methods taken from the Levison and Duncan's package SWIFT and RA15 Everhart's method.

3   Results

The obtained results were as follows:

- all of QB_1-like test particles (except 3) have very regular dynamical evolution over 200 - 300 million years,
- the results were method - independent,

- the regular evolutions shows that most of the particles have very stable orbits - this means that most of them will not be thrown away outside the KB in a time of million years (this is compatible with other researches), and
- of course, in this case it cannot be said that obtained solution is noised by numerical errors nor that the solution is natural.

Different behaviours can be observed for TL_{66}-like test particles. Evolutions of the same particles can run differently, dependent on the method or computer used in the integration process. That behaviour can be explained as an influence of deterministic chaos on the obtained solution. In that case the solution is treated as a one of the possible particle paths. But is it possible to be sure that solution is not just a numerical fluctuation ? This is very problematic question because chaos cannot be separated from the other types of numerical errors for general three- or more-body problems (there is no way to compare between numerical and an exact analytical solution).

4 Conclusions

All the conclusions can be summarized out as follows:

- numerical integration over long time-scales (several million of years) may be useful for quasi-circular orbits and when there are no close approaches between the test particle and a massive body,
- dynamical evolution of elliptic orbits is much more faster (due to approaches to massive bodies) and depends on two orbital parameters (eccentricity and inclination of orbit plane to the ecliptic plane),
- the length of the interval of time in which we compare the results, depends on the above mentioned two orbital parameters - the more elliptic the orbit and the less inclined, the shorter the interval of time,
- the length of time interval in which the integration gives comparable results can be estimated as 75 - 125 million years, and the minimum length to about 1 - 10 million years,
- a new investigation should be done to separate deterministic chaos from other types of numerical errors - otherwise it will not be obvious that the obtained results are scientific or are just a numerical error.

The values shown above are only estimates. They inform us only about the order of time interval of integration when the solution is supposed to be natural. The two last conclusions are not universal. If we have different n-body problem or test particles that are situated in other parts of the Solar System, we need to repeat the calculations to obtain desired maximum value of the time interval. Results of the calculations do not tell that there is no point to integrate over longer time scales. These results only show that you cannot do it in 64-bit precision and in the equatorial frame. If you use the 128-bit precision and choose ecliptic frame - then the maximum time interval is supposedly longer.

References

1. Duncan, M.J., Levison, H.F., Lee,M. H., 1997, AAS 191, 69.03,
2. Wisdom, J., Holman, M., 1992, AJ, 104, 2022,
3. Holman, M., Wisdom, J., 1993, AJ, 105,1987,
4. Levison, H.F., Duncan, M. J., 1997, Icarus, 127, 13,
5. Everhart, E., 1985, Dynamics of Comets: Their Origin and Evolution, 185.

Future Investigations with New Facilities

Hermann Boehnhardt

European Southern Observatory, Alonso de Cordova 3107, Santiago de Chile

Abstract. At the beginning of the third millenium more than 700 m² of mirror collecting power at 6-10m class telescopes combined with more than 20 sophisticated instruments will be available for astronomical research, and last but not least, also for the investigation of minor bodies in the outer solar system (MBOSS) from the ground. The detector and computer technology will support wide field applications at large to medium size telescopes. Telescopes operated in Earth orbit open access to almost all wavelength regions from the high energy to the coolest regimes in space. Large radio telescope arrays for the sub-mm wavelength region are in the planning phase. A spacecraft mission to Pluto/Charon and possibly to another Transneptunian object (TNO) appears over the horizon for future research.

The paper addresses the research possibilities with such equipment for various kinds of projects like: the search for new objects, the follow-up orbit assessment and the characterization of the physical and chemical properties of MBOSSs.

1 MBOSS 2000 and Beyond: The Main Questions

The exploration of the recently discovered members of the outer planetary system, the Transneptunian objects (TNO) and the Centaurs, is one of the most appealing and demanding subjects in solar system science. It is generally believed that the Transneptunian region is an important reservoir of primitive remnants from the formation period of the solar system, possibly even more primitive than the comets which received and still do receive a tremendous scientific interest by ground- and space-based programmes. Both object types, TNOs/Centaurs and comets, are considered to play a key role for the investigations focussing on the scenarios of the origin of our Sun, the planets and the solar system as a whole. Moreover, it is equally important to understand their evolution over the 4.6 billion years since their formation. This will finally allow to disentangle the traces of the origin from those of their long existence in the icy regions of the outer solar system.

Less than a decade after the discovery of the first TNO (1992 QB1; after Pluto/Charon) we are just at the beginning of more comprehensive and systematic studies of these objects. There are two questions which need to be answered in order to put TNOs and Centaurs into the global picture of our solar system and to relate them to other groups of bodies therein:

1. the orbits and the dynamics or in other words the distribution in space and time

2. the object and group characteristics or in other words the physical nature (size, shape, density, mass, albedo, rotation, chemistry, surface/internal activity, surface layer stratification and internal structure etc.)

Although individual objects can represent interesting and challenging cases for our understanding of TNOs, it is important to ascertain the answers to the above mentioned question through systematic and very accurate studies of a statistically significant and representative sample. The sample size needed is not easy to estimate. However, it is difficult to imagine that major conclusions can be based upon a sample of much less than 100 objects (i.e. slightly more than the number of presently known TNOs; for the Centaurs the situation is worse).

The future observational studies ("future" in this context means roughly within the coming 5 years) will utilize existing and new technologies and methods for the observations and analysis of these objects. In the optical wavelength range the workload will be distributed among the medium-size (2-5m aperture) and the new generation of large telescopes (8-10m aperture). New instrumentation will allow to address so far unaccessible subjects by observations. Far-IR and sub-mm telescopes are at the limit for detection of a TNO signal in the long wavelength range. And after all, even a space mission will be launched which will visit after the flyby at the binary TNO Pluto/Charon yet another (still undiscovered) TNO further away from the Sun.

2   Wavelength Regions for Remote Detection: Visible, Near-IR and Sub-mm Windows

All TNOs and Centaurs known so far were first observed in the visible wavelength range. Although the surface albedo of the objects is not known (most likely between the albedo of Pluto - 0.55 - and of comets - 0.04 or less), they are clearly detectable and measurable in the visible and near-IR region.

The equilibrium temperature at the surface of an object at about 40 AU from the Sun is very low (Encrenaz et al. [7]) since the solar illumination is weak (neglecting internal heat sources like radioactivity and recrystallization; Prialnik [15]): between 40 and 50 K depending on the albedo and rotation period. Therefore, the thermal emission of the object will be in the sub-mm wavelength range with flux maximum between 50 and 100 μm. For Centaurs the surface temperature is expected to be - at best - 20-30 K warmer and their thermal emission thus peaks at about 40 μm. For the flux of MBOSS objects in the optical and radio wavelength region see Encrenaz et al. [7].

Considering the technology (telescopes and instruments) which is currently available and which will be available within the next few years, clear advantages of TNO and Centaur exploration (in terms of detectibility) exist in the visible and near- IR wavelength range as compared to the sub-mm region. This applies to both imaging and spectroscopic observations.

3   The Classical Approach: Observations in the Visible and Near-IR from Earth

The investigation of TNOs and Centaurs in the visible and near-IR wavelength range from Earth concentrates on the search for MBOSS and on the exploration of the nature of the objects, i.e. on the population statistics of the orbits and the basic physical properties. For both issues 2-5m class telescopes work at the limit of their capabilities and, in particular for the physical characterization of the populations, 8-10m class telescopes are needed.

3.1   MBOSS searches: orbit statistics and population

Search projects for TNOs aim to be wide and/or to be deep. As one can see below, these search techniques have (or will have) the adequate telescopes and instrumentation available by now (or soon). And they will be complementary in their research goals, although both search methods are needed to achieve a reasonably complete picture of the object population in the outer solar system. For all search methods the object recognition relies on the motion of the object, i.e. it invokes either long exposures which allow to identify the moving objects by their trails (difficult for TNOs because of their short trail lengths due to the slow motion rate) or by the position changes of the objects between multiple exposure of the same sky region.

Wide-field searches. Efficient TNO searches over a larger area on the sky require sensitive wide field-of-view telescopes/instruments (the classical photographic surveys with Schmidt telescopes are wide, but do not go deep enough). Depending on the target distance and expected limiting magnitude careful considerations of the observing strategies for wide angle coverage are needed: the simple approach of two subsequently taken images may not be adequate for all purposes.

Over the past few years the search for TNOs was performed in the visible wavelength range either at 2-5m telescopes with narrow field-of-view cameras (Jewitt et al. [13]; Gladman et al. [11]) or at smaller telescopes (about 1m aperture) with wider fields-of-view, but with less sensivity. However, with the recent improvements in CCD technology (large CCD array mosaics of 8×8k pixels or more) the suitable equipment is or will soon be installed at several medium-size telescopes. The following compilation summarizes the basic telescope and instrument specifications, where to find such a telescope and what science can (and cannot) be addressed (naturally, this list may be incomplete).

Telescope/instrument properties:

- 2-5m telescopes

- imaging instruments with 0.5 - 1 deg field of view (or more) for broadband filters
- large CCD mosaics (8×8k to 16×16k pixels) with high pixel resolution and fast, but low-noise read-out electronics
- 24-25mag limiting brightness

Sites:

Observatory	Telescope	Instrument	Status
Kitt Peak	4m tel.	Mosaic Imager	operational
La Palma	2.5m INT	wide field camera	operational
La Silla	2.2m MPG/ESO	WFI	operational
Hawaii	3.5m CFHT	MEGACAM	in commissioning
Tololo	4m Blanco	MOSAIC II	in commissioning
Paranal	2.5m VST	OMEGACAM	installation in 2001

Science:

- efficient wide angle searches for TNOs and Centaurs: in and out of ecliptic, with the aim to cover the ecliptic band almost completely
- orbit characterization of all detected objects: safe recoveries and follow-up observations
- orbit population statistics: resonances and dynamical TNO families of the inner Edgeworth-Kuiper Belt (EKB)
- inventory of large bodies (> 100 km) in inner EKB
- search for other large Pluto-type TNOs

Wide searches with 2-5m class telescopes require good atmospheric conditions (dark sky with seeing below 1 arcsec) and enough observing time to achieve the optimum coverage on the sky and to guarantee safe recoveries of the objects for the orbit determination. A synergetic and challenging side effect of such wide searches for the understanding of the formation and development history of the outer solar system can result from the combination of the orbit statistics with the physical classifications of the objects in the EKB. The detection limit of 24-25mag mentioned above corresponds to typical exposure times of about 30 min for medium size telescopes under good seeing. It clearly limits the distance range for TNO detections (good chances most likely only for the inner EKB) and it will also not allow to fully explore the size distribution of the objects (the smaller ones will be missed).

Deep pencil-beam searches. Deep pencil-beam searches for TNOs are typically performed with the most sensitive telescopes and instruments, i.e. large telescopes (on the ground: 4-10m aperture; in space: HST) with efficient imaging systems. So far, the limiting factor for pencil-beam searches is the narrow field-of-view of the available imaging systems. The characteristics of a pencil-beam search is that it collects a series of images at the same region of the sky and combines these exposures for the detection of very faint moving objects in the (narrow) field of view.

Such searches for TNOs are not new. However, the past attempts were either not deep enough ($\geq$ 26 mag) to sense the small-size objects and/or the larger, but more distant ones (see for instance the 5m Palomar telescope search by Gladman et al. [11] and the 3.5m ESO NTT search by Boehnhardt et al. [2]) or they led to uncertain and unverified detections of objects (see the HST search by Cochran et al. [5]).

8-10m class telescopes and the HST can go deeper and are thus more adequate for pencil-beam searches of TNOs. Hence, a detection limit of 30mag (through a broadband filter) is not out of range (FORS@VLT reaches this magnitude in 12h total integration time for 0.4 arcsec seeing). According to model prediction for the population of the EKB the spatial density of TNOs is high enough to make the detection of objects in narrow fields very likely if the limiting magnitude is 27 mag or fainter (Jewitt et al. [13]).

Telescope/instrument properties:

- 8-10m telescopes or HST
- imaging instruments with 5-10 arcmin field of view for broadband filters
- normal size CCDs ($2 \times 2k$ to $4 \times 4k$ pixels with very good pixel resolution)
- about 30mag limiting brightness

Sites:

Observatory	Telescope	Instrument	Status
Earth Orbit	HST	WFPC	operational
Paranal	VLT	FORS	operational
Hawaii	KECK	LRIS	operational
Hawaii	SUBARU	FOCAS	installation in 1999
Hawaii	GEMINI North	GMOS	installation in 2000
Pachon	GEMINI South	GMOS	installation in 2001
Mt. Graham	LBT	optical imager	installation in 2002

Science:

- detection of distant EKB objects beyond 100 AU
- small TNO population in inner EKB (< 100km)
- tomographic sensing of EKB versus solar distance and object radius

Pencil-beam searches require the best image quality and seeing conditions in order to go really deep. Because of the limited telescope time available for such programmes at the large facilities they may only be done in the most sensitive broadband filter which will reduce the ability of the physical characterization of the object. For the same reason any follow-up observations after the detection of a very distant and faint TNO may be difficult since no telescope time may be scheduled and the object may be too faint for follow-up observations at 4m class telescopes. Therefore, the orbit determination will become problematic and the detected objects may become lost again if not re-observed within a certain time interval.

The ultimate TNO search challenge: wide and deep. Within the next years it will be possible to combine the benefits of both observing methods by using the proper telescope/instruments for a wide (around 0.5 deg field of view) and deep (28-30mag) TNO search. With the advent of prime focus and wide field Nasmyth cameras at large telescopes new limits for the global exploration of the outer solar system beyond 100 AU can be reached.

Sites:

Observatory	Telescope	Instrument	Status
Hawaii	SUBARU	SUPRIME-CAM	install. 1999
Paranal	VLT	VIMOS	install. 2000
Mt. Graham	LBT	prime focus imager	install. 2002 (?)

Data analysis aspects. The search for very distant and faint moving objects in the outer solar system requires special attention and effort for the data analysis. The wide field imagers produce large amount of data per observing run (for instance: about 130MB per FITS image with the La Silla WFI instrument or about 30 GB per night) which have to be processed at appropriate computer platforms. The object detection in deep pencil-beam searches will have to struggle with the crowded background of galaxies. Special image processing techniques (see for example Boehnhardt et al. [1], Hainaut et al. [12]) can be useful to remove the background objects self-consistently and non-destructively for the potential TNOs in the images. Finally, both search approaches deserve new methods to support the automatic detection of moving objects in the data (for instance using the SEXTRACTOR tool). The painful way of visual blinking of the images - as applied for TNO detections by many observers - is by far not adequate to deal with the complexity and the large amount of data expected from the wide field and deep pencil-beam TNO search projects.

3.2 Physical studies

The physical studies aim for the establishment of a type classification scheme for MBOSS objects. With the sparse observational results obtained over the past few years (mostly at 3-5m class telescopes) we are just at the beginning of this process which in some sense has analogies to the same efforts undertaken three decades ago for the main belt asteroids. The observational challenge for TNOs and Centaurs lies in their faintness which makes it much more difficult than for asteroids to measure their size, shape and rotation, the colours, spectral taxonomy and polarimetric properties. From the existing results for Pluto (see de Bergh [3] and references therein) as well as for a few TNOs and Centaurs (see Davies [6] and references therein) it is expected that the near-IR region (in particular the H and K bands) may provide the best chances to distinguish objects by their colours and to identify spectroscopically absorption bands from surface materials (best candidates are CO, CO_2, CH_4 and N_2 ices). Because of the hypothesized relationship between EKB objects and short-period comets (Levison and Duncan [14]) and since the equilibrium temperature of the bodies at 20-50 AU may still be high enough to support the sublimation of surface ice, a search for a coma-like atmosphere (for larger bodies even a permanent atmosphere may be possible) around TNOs and Centaurs may not be a hopeless goal for the observations. If successful (the first marginal detection of a coma around 1994TB was announced during this workshop; Fletcher et al. [9]), it will provide an important link between both classes of solar system bodies. However, the measurement of the albedo of the objects requires a different approach both in observing technique (see section 4.1) and wavelength range (see section 5).

The final goal for these studies is to identify (in combination with the orbit statistics) the role of the MBOSS objects in the formation scenario of the solar system and to understand the development of these bodies with time and in space.

Size, Shape, Rotation and Colours: these physical parameters of TNOs and Centaurs are assessed through imaging observations in the visible wavelength range. For the large and closest objects (brightness between 20-25mag) 3-5m class telescopes will do the job. As discussed during this workshop, there seems to be an inconsistency of the colour photometry obtained for the same TNOs with different instrumentation which deserves clarification by a reanalysis of the data (see the discussion by Davies et al. [6] in these proceedings). An elegant and fast way to compile a fairly complete and consistent set of broadband photometric measurements of the TNOs and Centaurs could be a few dedicated runs at a large (8-10m class) telescope during which basically all known objects could be measured over a year (the estimated observing time is of the order of 4-5 nights).

For the smaller and more distant (both still to be detected) TNOs with expected brightness between 25 and 28 mag even the photometry has be performed at 8-10m class telescopes or the HST in order to achieve the proper accuracy (a few hundredth mag). The same applies for the detection of a - likely low surface brightness - coma activity which in addition requires high image definition (about 0.1 arcsec/pixel) and excellent seeing conditions in order to resolve the brighter central part of the atmosphere.

For the availability of adequate 8-10m class telescopes and instrumentation for such programmes, see the list in section 3.1.

Taxonomy: these studies need photometric and low-dispersion spectroscopic observations of the objects in the visible and in particular in the near-

IR wavelenght range. Generally, spectroscopy can be obtained - with the required accuracy of S/N > 10 - for the brighter TNOs and Centaurs (< 25 mag) only at the largest telescopes in the world (on ground and in space). As mentioned above, spectroscopy in the near-IR wavelength range may be the most interesting tool to sense the surface material of the bodies.

Site:

Observatory	Telescope	Instrument visible	Status
Earth Orbit	HST	WFPC,STIS	operat.
Paranal	VLT	FORS	operat.
Hawaii	KECK	LRIS	operat.
Hawaii	SUBARU	FOCAS	install. 1999
Hawaii	GEMINI North	GMOS	install. 2000
Pachon	GEMINI South	GMOS	install. 2001
Mt. Graham	LBT	Spectrograph	install. 2002
McDonald	HET	LRS	install. 1999

Observatory	Telescope	Instrument near-IR	Status
Earth Orbit	HST	NICMOS	operat.
Paranal	VLT	ISAAC; CONICA	operat.; install. 2000
Hawaii	KECK	NIRC	operat.
Hawaii	SUBARU	CISCO; CIAO	install. 1999; 2000
Hawaii	GEMINI North	NIRI; GAOS	install. 2000
Mt. Graham	LBT	near-IR instr.	install. 2002

Science (see also above):

- colour classification of small and distant TNOs and Centaurs (visible and near-IR)
- surface chemistry and taxonomy: detection of ice absorptions
- population types (for instance versus solar distance and/or size)

The following table lists the detection limits (in mag) for FORS, ISAAC and CONICA (including the adaptive optics system NAOS) at the ESO VLT in order to illustrate the capabilities of modern instrumentation at an 8m telescope.

Wavel. range	FORS (imag.)	FORS (spec.)	Wavel. range	ISAAC (imag.)	ISAAC (spec.)	CONICA (imag.)	CONICA(spec.) (spec.)
U	27.0	21.9	J	24	21	26.9	24.0
B	27.9	23.4	H	23	20	26.3	23.5
V	27.2	23.2	K	22	19.5	25.4	22.6
R	26.6	22.8	L	18	15.5	19.1	16.7
I	26.3	22.5	M	15	13	17.1	14.5

Remarks: the following technical parameters for the observations are as-sumed:

FORS(ima.): exposure time = 1800 sec, S/N = 5, seeing = 0.5 arcsec

FORS(spec.): exposure time = 1800 sec, S/N = 10, slit width = 1 arcsec, spectral resolution $\sim$ 700

ISAAC(ima.): exposure time = 3600 sec, S/N = 5, seeing = 0.5 arcsec

ISAAC(spec.): exposure time = 3600 sec, S/N = 5, slit width = 1 arcsec, spectral resolution $\sim$ 500

CONICA(ima.): exposure time = 3600 sec, S/N = 3, with adaptive optics system NAOS

CONICA(spec.): exposure time = 3600 sec, S/N = 3, spectral resolution $\sim$ 500, with adaptive optics system NAOS

It is noteworthy that MBOSS science will benefit considerably from near-IR instrumentation with adaptive optics (AO) systems (like CONICA/NAOS at the VLT or similar equipment at other telescopes). The gain in limiting brightness as compared to normal near-IR instruments is about 3-4 mag. The use of natural reference stars for the image analysis of the AO system will restrict the application of this type of observations to a few special cases. However, when Laser guide star systems become available in addition to AO instruments at 8-10m class telescopes, the near-IR range will represent the best and most powerful wavelength region for physical/taxonomic studies of TNOs.

Polarimetric studies: several of the above mentioned instuments (like FORS at the VLT) also provide options for polarimetric measurements (both in imaging and spectroscopy mode) which could be applied for TNO and Centaur observations. However, since these objects will be observed under rather small phase angles ($< 10°$), only a very small part of the polarisation curve (and most likely the one with the smallest polarisation degree) can be measured. Therefore, the impact of polarimetric studies on the physical characterization of TNOs may be limited.

4   MBOSS Observations with Special Means

This section describes three types of observations which use special options and configurations of ground-based optical telescopes for the measurement of physical parameters of very distant solar system bodies which are otherwise very difficult or impossible to estimate: the shape and albedo, the binaries or satellite systems and the population of small bodies down to a few 10m size.

4.1   Albedo, size and shape by adaptive optics
and optical interferometry

The physical parameters "size and shape" of TNOs and Centaurs become
directly measurable through high-definition imaging which allows the spatial resolution in the milliarcsec range. Using the size estimation from high-
resolution imaging and accurate photometry of the body measured (simultaneously) in the same (or similar) wavelength range, the surface albedo can
be obtained. In that way time-series observations of the lightcurve variability
will be disentangled into shape and albedo contributions. For TNOs the goal
is to resolve a 20-22mag object of typically 100km size at a distance of about
50AU, i.e. about 0.0025 arcsec diameter. For 15-17mag Centaurs of 500km
size the resolution scale is 0.05 arcsec at 15AU.

Adaptive optics systems (AO): AO systems like CONICA+NAOS at the
VLT will provide spatial resolution of about 0.03 arcsec in J band down to
26mag (using natural or Laser guide stars) which is of the order of the resolv-
ing power of the HST imaging mode. At best, both AO and HST imaging
will be able to measure directly the dimensions and albedo of very large and
"close" Centaurs which, however, still need to be discovered.

Optical interferometry (OI): OI system will achieve higher spatial resolution as compared to AO instrumentation. The interferometric mode of
the VLT (VLTI) will support imaging of 0.001 arcsec resolution at 1 μm.
VLTI will work in the visible, near- to mid-IR wavelength range and will
also provide a spectroscopic mode (from 0.5 - 25 μm). In K band a limiting
brightness of 23 mag is expected for imaging (while a reference source of
< 16mag is needed for the parallel image analysis). Therefore, OI systems
like AMPER at the VLTI (installation in 2001) should allow to resolve large
TNOs and Centaurs in the near-IR wavelength range and to measure the basic object properties size, shape and albedo directly. Such results will provide
new insights in the surface structure and body shape of TNOs and Centaurs.

4.2   Binary systems

The Pluto/Charon systems is a TNO binary the origin of which is presently
unknown. Also among the asteroids double systems (main component plus
a satellite) are found (asteroid 951 Gaspra is a very prominent case detected by the Galileo spacecraft). The question may be allowed whether the
Pluto/Charon system is unique in the EKB region. For bound binaries with
components of planetesimal nature (like TNOs) the gravitational domain may
extend from 1000 - 20000 km or 0.03 - 0.5 arcsec on the sky depending on
the mass of the system.

High-definition visible and near-IR imagers will be able to resolve the
wide TNO binaries under good seeing conditions. AO and OI systems at
large telescopes are the much better choice for the closer and even the tough
cases of binary TNOs (however, with limitations for the object magnitude).

Compared to these options the classical approach of occultation lightcurves is much less promising, since it will only detect very special cases of the likely systems - if any (i.e. those with occultations for an observer on Earth) - and it requires lots of observing time to identify the interesting candidates.

The scientific value of the detection of TNO binaries lies in the possibility for the mass and density estimation of the objects and - if more cases are found - in the implications for our scenarios of the planetesimal formations and their dynamical stability.

4.3 Star occultations

The occultation of stars by objects in the outer solar system may occur if there are many of them (both stars and TNOs) in a given field of view. Close to opposition the duration of an occultation of a point-like stellar source by a TNO of 100km diameter is only just 3 sec. This duration is basically independent from the solar distance of the moving object (the occultation duration $\delta T_{occ} \sim 9 \times 10^{-6} \times D$ with δT_{occ} in hours and the diameter D in km). However, close to the stagnation points of the annual visibility window the occultation durations become much longer (due to the slower motion rate of the object) and will thus allow to detect also very small TNOs. The elongation of the stagnation points reflect approximately the distance of the occulting object from the Sun. For a more detailed discussion of this method and the expected results see the paper of Roques [16] in these workshop proceedings.

The technical specifications for such occultation experiments are:

- high-speed imaging or photometer system
- time resolution (0.1 sec or better)
- many stars in field of view
- several (small) telescopes

The science return of successful occultation observations of TNOs will be:

- dimensions of small to large TNOs (lower limits only)
- number statistics for the small EKB population

Although the position of the occulted star can be measured very accurately, there may be no chance to determine the orbits of the moving - and usually extremely faint - objects. Also any kind of further information on the physical nature of the occulting body cannot be extracted from such measurements.

Beside the satellite project described in Roques [16] there is also the planning of a ground-based multi-telescope observatory (several half-metre size telescopes) underway in Taiwan (project TAOS, a joint venture between institutes in Taiwan and the United States).

5   Far Infrared and Sub-mm Observations

As outlined in section 2, the far-IR and sub-mm wavelength range represents the second window which could be used for observations of very distant
objects in the solar system. The scientific goals for such observations are:

- the thermal budget of the icy bodies
- the surface albedo determination
- the detection of an atmosphere and molecules therein

The thermal budget and albedo can be determined (in addition using
simultaneous flux measurements of the reflected sunlight from the object
in the visible and/or near-IR wavelength range) if the thermal continuum
radiation of the body is detected. The maximum thermal emission is expected
around 100 μm, so both far-IR telescopes and sub-mm antennas may be
suitable for such observations.

The detection of a coma or bound atmosphere around a TNO or Centaur
is best focused on the CO rotational transitions (2-1, 3-2, 1-0) in the sub-mm
wavelength range. Theoretical models (see Prialnik [15] and Delsemme [4])
predict that CO ice can sublimate well beyond 40 AU from the Sun although
the production rate may be very low. Internal heat sources (even temporary
ones, see Prialnik [15]) can contribute to the surface ice sublimation. Based
upon the CO productions rates of comet Hale-Bopp measured with the ESO
SEST antenna at La Silla when the comet was at about 7 AU outbound
and applying simple scaling laws for the CO sublimation rates versus solar
distance, one can expect that a TNO with a CO atmosphere can produce a
CO emission flux which is a factor of 1000 smaller than the Hale-Bopp CO
production at 7 AU. From observations of long-periodic comets far from the
Sun it is known that a dust coma can still be present at solar distances of
20-30 AU. One can assume that a driving gas accelerates the dust into the
cometary coma. Possible candidates for driving agents at 30-40 AU from the
Sun are sublimating ices of: CO, N_2 and CH_4.

The actual status: the ISO satellite allowed the first likely detection of
thermal radiations from distant solar system objects (Thomas [17]). Sub-mm
measurements are not yet reported.

Which facilities are or will be available for observing programmes of TNOs
and Centaurs in the far-IR and sub-mm wavelength range?

The far-IR - SIRTF: NASA's Space Infrared Telescope Facility mission
SIRTF will carry a 0.85m size telescope with instruments for near-, mid- and
far-IR observations into Earth orbit. The launch of the satellite is planned
for 2001. From the on-board experiments in particular the MIPS Multiband
Imaging Photometer will be of interest for observing objects in the outer
solar system. MIPS will have detections channels at 24 μm, 70 μm and 160
μm and will be more sensitive than the ISOPHOT instrument by a factor of
at least 10 which makes it a very suitable tool for the detection of thermal
radiation from TNOs and Centaurs.

Sub-mm telescopes: based upon sensitivity considerations the sub-mm antennas of the SEST, JCMT and CSO telescopes may not be sufficient for a successful detection of CO emission from TNOs. IRAM (namely Plateau de Bure) and Owens Valley Telescope may have a chance for very long integration times. The success rate of such observations will be high for the ALMA facility currently planned as a joint project between ESO and United States institutes (see section 7). Of great help for the selection of the targets would be the guidance by (likely) coma detections around objects in the visible wavelength range.

6 Space Mission: Pluto Express

NASA considers a proposal to visit the Pluto/Charon system by a scientific probe in order to perform the in-situ exploration of this unique binary in the outer solar system. The launch of the mission (called Pluto Express) is proposed for early in the first decade of the next millennium with an arrival time at Pluto/Charon about 10 years later. The scientific goals are:

- the characterization of the global geology and geomorphology of Pluto and Charon
- the mapping of the surface composition
- the characterization of the neutral atmosphere in terms of composition, thermal structure and aerosols

The spacecraft will carry three scientific experiments:

- a high resolution imager
- an IR spectral mapper
- an UV spectrometer

Since Pluto and its companion Charon possibly represent a special and so far unique binary system of TNOs, one can expect that such a mission, if approved and successfully performed, will revolutionerize our knowledge and understanding of the icy bodies in the outer solar system and the EKB. As an add-on Pluto Express can be targeted to a more distant TNO after the fly-by at Pluto/Charon and it can repeat the experiments at an EKB object beyond Pluto (in that way becoming the Pluto-Kuiper-Belt Express Mission).

Up to now, no firm decision about the Pluto Express mission has been made. For the extended mission into the EKB the spacecraft is still lagging a suitable target object. However, such an object is likely to exist considering the estimations of the expected number of TNOs in the EKB reservoir. The selection and decision on the target has no rush, since the path of the spaceprobe can still be adapted to the Transpluto TNO by a suitable maneuvre which will be executed mid-course on its way to the prime target.

The potential targets for the extended mission have to be searched for in an region of several 10 deg around the position of Pluto on the sky. Therefore, the best approach for finding the targets is a wide angle search for TNOs in the vicinity of Pluto's coordinates on the sky. The only serious difficulty for such a search programme will be the fact that Pluto is in front of star-rich Milky Way regions now and for a long time to come which will make the detection of the likely rather faint TNOs a delicate task (see the discussion about the galaxy background for the deep pencil-beam searches in section 3.1). Furthermore, it will be even more difficult to characterize the potential targets for their physical properties since even with the largest telescopes the normal recipes for such observations (as described in section 3.2) will easily be spoiled by the crowded background and very careful considerations on the planning of the observations and on the observing techniques have to be made.

7   ALMA, NGST and OWL: An Outlook

The sections above outline the scientific investigations of minor bodies in the outer solar system using new and future facilities on Earth and in space. The time span during which most of this MBOSS research will be performed or brought on its way is most likely of the order of 5 years from now (beginning of 1999).

However, a few of the scientific goals defined above may still be left open by then and some new big questions in this area of research may have shown up which will need even more powerful and sophisticated telescopes to be answered. For instance, the population of small planetesimals in the EKB and extremely distant solar system objects (beyond 500-1000 AU if they exist) may still be widely unexplored.

Which telescopes projects could one think of to be used in the somewhat more distant future (let's say in 10 years from now) for the investigation of the still unanswered problems in the MBOSS research?

ALMA, the Atacama Large Millimeter Array: ALMA is a joint project between the European Southern Observatory ESO and American institutes to build a large mm/sub-mm telescope array in the Chilean Atacama desert. In the final installation this array will consist of 64 12m antennas. As already mentioned in section 5 it will be suited to attempt - with a chance of success - measurements of the thermal continuum emission of at least the largest TNOs and will have much better capabilities for the detection of gas emission and atmospheres around these objects.

NGST, the New Generation Space Telescope: this 6-8m aperture space telescope is NASAs new project in continuation of the very successful HST. The NGST will contribute best to the near-IR based investigations of TNOs and Centaurs and the physical characterization of the smaller population of MBOSS objects will be a very interesting application for this telescope.

OWL, the Overwhelmingly Large Telescope: OWL (Gilmozzi et al. [10]) is a concept for a 100m telescope currently studied by ESO. Equipped with adaptive optics this telescope will reach faint objects down to 38 mag in 10h total integration time. This will allow the detection of TNOs with 200 km radius at about 2000AU or Hale-Bopp size bodies (25 km radius) to almost 600 AU. In other terms, this telescope will allow to study the EKB to its expected very edge of extension and to observe large and smaller objects therein. It may even be possible to detect the hypothesized outer ring of icy bodies beyond the EKB (Fernandez and Ip [8] and references therein).

References

1. Boehnhardt, H., Babion, J., West, R.M., 1997, A&A 320, 642
2. Boehnhardt, H., Hainaut, O., Delahodde, C., West, R.M., Meech, K., Marsden, B., 2000. In these proceedings.
3. de Bergh, C., 1998. Presented at this Workshop.
4. Delsemme, A., 1982, in: Comets, ed. L.L. Wilkening, Univ. Arizona Press, Tucson, 85
5. Cochran, A.L., Levison, H.F., Stern, S.A., Duncan, M.J., 1995, ApJ 455, 342
6. Davies, J.K., 2000. In these Proceedings.
7. Encrenaz, T., Bibring, J.-P., Blau, M., 1991, The Solar System, Springer Press, Heidelberg, chapter 1.2.1
8. Fernandez, J.A., Ip, W.-H., 1991, in: Comets in the Post-Halley Era Vol. 1, eds. Newburn, R.L., Rahe, J., Kluwer Academic Publishers, Dordrecht, 487
9. Fletcher, E., Fitzsimmons, A., Williams, I.P., Thomas, N., Ip, W.-H., 1998. Presented at this Workshop.
10. Gilmozzi, R., Delabre, B., Dierickx, P., Hubin, N., Koch, F., Monnet, G., Quattri, M., Rigaut, F., Wilson, R.N., 1998, Proc. SPIE 3352, 778
11. Gladman, B., Kavelaars, J.J., 1997, A&A 317, L35
12. Hainaut, O., Meech, K.J., Boehnhardt, H., West, R.M., 1998, A&A 333, 746
13. Jewitt, D., Luu, J., Chen, J., 1996, AJ 112, 1225
14. Levison, H.F., Duncan, M.J., 1997, Icarus 127, 13
15. Prialnik, D., 2000. In these proceedings.
16. Roques, F., 2000. In these proceedings.
17. Thomas, N., Eggers, S., Ip, W.-H., Lichtenberg, G., Fitzsimmons, A., Keller, H.U., Williams, I.P., Hahn, G., Rauer, H. 2000. In these proceedings.

A SUBARU Survey Project with Suprime-Cam

Jun-ichi Watanabe

National Astronomical Observatory, Osawa, Mitaka, Tokyo, 181-8588, Japan

Abstract. The SUBARU is one of the most appropriate telescopes for survey work among large telescopes because of the ability of wide field imaging at the prime-focus by using the Mosaic CCD camera, Suprime-Cam. We are planning to perform a survey of the Edgeworth-Kuiper Belt Objects as a long-term project for the SUBARU telescope. Based on the theoretical predictions, the expected discovery rate of new objects in the Kuiper Belt is also noted in this short report.

1 Introduction

One of the important priorities of astronomy is to establish the concept of the structure of our universe. The mapping of distant galaxies helps us understand the large scale structure along with its history. This is also the case in the solar system, where we are living now. The innovation of astronomical instruments allows us to see more distant structures of the solar system.

Astronomical telescopes changed the concept of the solar system from the 17th century. In 1781, F. W. Herschel discovered Uranus by using his telescope. This discovery made people recognize the existence of the outer structure beyond the Saturn. The development of the celestial mechanics allowed some astronomers to predict outer planets, which resulted in the discovery of Neptune in 1846. Further development of the photographic technique allowed C. Tombaugh to discover Pluto. However, probing more distant places could not be realized until the invention of CCD detectors. In 1992, D. Jewitt and J. Luu (1993) discovered a small object, 1992 QB$_1$, outside Pluto's orbit by using CCD.

More than seventy objects have been found until now (end of 1998) in this outer solar system region. These objects are thought to be members of the Edgeworth-Kuiper Belt, which was originally predicted by K. Edgeworth in 1943, and by G. Kuiper in 1950. Recent theoretical studies of Duncan et al. (1988) indicate that it should provide a source of short period comets. Holman and Wisdom (1993) estimated the total mass of this belt would be 0.02 to 0.2 of Earth's mass, suggesting the existence of huge objects in this region beyond Pluto. More sophisticated observations are needed to understand the Edgeworth-Kuiper Belt. In this paper, we point out that the SUBARU is an appropriate large telescope for this purpose, and propose the survey of the minor bodies in the outer solar system as a long-term project for the SUBARU telescope. This survey will serve to determine the structure of our solar system.

2   Why the SUBARU Telescope?

The SUBARU, named after the old Japanese word for the open cluster Plei-
ades, is the first national large telescope of Japanese astronomical community.
It has a 8.2-m mirror(F/2) of thin-meniscus type, and is located on Mauna
Kea, Hawaii. The telescope had its first light in early 1999.

There are two important reasons for performing the survey work by the
SUBARU. One is the depth of the limiting magnitudes. Because most of
the Edgeworth-Kuiper Belt Objects are expected to be fainter than 23rd
magnitude, we need large telescopes to detect new objects. When we see the
Edgeworth-Kuiper Belt objects of 1992 QB_1-size at 80 A.U., which is about
two times farther than the 1992 QB_1 itself, the magnitude would be expected
to be 26. This magnitude can be reached by the new 8-m class telescopes.

The other reason for a SUBARU survey is the wide field of view. The
wide field of view would be generally realized at the prime focus, which most
of the planned 8-m class telescopes will not have except the SUBARU and
the LBT. Therefore, the SUBARU telescope is one of the most appropriate
telescope projects due to the ability of wide field imaging at the prime-focus
by using the Mosaic CCD camera, which is called as Suprime-Cam. The
expected field of view of the Suprime-Cam at the prime focus of the SUBARU
is 30×24 arcmin^2, which will be covered by 10 CCDs of each $2K \times 4K$
pixels. The limiting magnitude in the R-band is 26.6 for S/N = 10 with one
hour integration. The first generation of such Mosaic CCD camera has been
developed so far (Sekiguchi et al. 1992), and works well at the prime focus
of the 105-cm Schmidt telescope at the Kiso Observatory. The Suprime-Cam
has been operational at the prime focus of SUBARU since the autumn of
1999.

3   Expected Discovery Rate of the Edgeworth–Kuiper Belt Objects

Although we do not have enough objects to discuss the population or structure so far, there are some theories to be used to estimate the discovery rate
of the Edgeworth-Kuiper Belt objects with the Suprime-Cam. Yamamoto
et al. (1994) suggested that both the orbital properties and estimated size
are consistent with those of the remnants of the planetesimals formed at the
beginning of the solar system. Although there is some uncertainty in the parameters used, the number density of objects has a maximum at around 100
A.U. to 200 A.U. This suggests that the discovered objects around 40 A.U. so
far is only tiny population of the inner part of the outer structure. Moreover,
Holman and Wisdom (1993) estimated the total mass of the Kuiper Belt as
0.02-0.2 Earth's mass.

The important point revealed from these theories is that there are more
objects at more distant places. Yamamoto et al. (1994) predicted the spatial

distribution along with the cumulative number of the remnant planetesimals. Combining the total mass predicted by Holman and Wisdom (1993), we w estimated the discovery rate of new objects in the outer region on the basis of the theory of Yamamoto et al. (1994) when we use the Suprime-Cam at the prime focus of the SUBARU telescope. The result is nearly one object per one exposure. This discovery rate is by order of magnitude higher than the classical 2-m class telescopes. Although we assumed the limiting magnitudes of the SUBARU is 26 in this estimate, much fainter magnitude can be attained when we apply a new technique such as the shift-and-add method.

We recognize how efficiently we can detect new objects using the SUBARU telescope. That is the main reason why we are proposing the survey of the Edgeworth-Kuiper Belt objects as a long-term project for the SUBARU telescope.

References

1. Duncan, M., Quinn, T., Tremaine, S., 1988, *Astrophys. J. Lett.*, **328**, L69.
2. Holman, M.J., and Wisdom, J., 1993, *Astron. J.*, **105**, 1987.
3. Jewitt, D., and Luu, J., 1993, *Nature*, **362**, 730.
4. Sekiguchi, M., Iwashita, H., Doi, M., Kashikawa, N., and Okamura, S., 1992, *Publ. Astron. Soc. Pacific*, **104**, 744.
5. Yamamoto, T., Mizutani, H., and Kadota, A., 1994, *Publ. Astron. Soc. Japan*, **46**, L5.

The PICOCAM Project at Pic du Midi

François Colas[1], Jean Eudes Arlot[1], Jerome Berthier[1], Agnes Fienga[1],
Michael Gastineau[1], Daniel Hestroffer[1], Laurent Jorda[2], Jean Lecacheux[3]

[1] Institut de mécanique céleste/BDL, URA 707 du CNRS, 77 avenue Denfert
Rochereau, F-75014 Paris, France
[2] Max-Planck Institut fur Aeronomie, Lindau, Germany
[3] Observatoire de Meudon, DESPA, Meudon, France

Abstract. Detection of minor bodies in the outer solar system depends of course
on the telescope aperture but also on the field-of-view. Our idea was to use a small
55-cm telescope at the Pic du Midi Observatory that was initially designed for
photometry. We took advantage of the small size of modern CCD cameras to use
the prime focus of the F/3 mirror. We designed a three lens corrector to obtain a
one degree square field. In this configuration we hope to scan 10 000 square degrees
up to m = 21. If we use a standard luminosity function (Jewitt et al., 1998), we
hope to detect 500 Centaur or Kuiper Belt Objets. Another use of this automated
telescope can be an exhaustive survey (photometric and astrometric) of the existing
comets brighter than magnitude 20.

1 Description of PICOCAM

We used a 55-cm telescope previously designed for stellar photometry. Our
main goal was to build a fully automated telescope that may be used in
remote control mode. To obtain good results, the telescope must work all the
year with the same observational strategy.

We also made a special effort to maximise the observational efficiency.
For this purpose we changed the mechanical design of all the motions. The
first thing was to speed up the pointing velocity to avoid important loss of
time between two fields. We obtained a pointing velocity of about 6 degrees
per second. As the survey strategy is to observe two adjacent regions, the
loss during the pointing procedure is small compared with the read out CCD
time of 10 seconds.

The second thing is the loss of time during the research of a guiding
star. As the mirror has a fast aperture (F/3), the maximum exposure time is
about 5 minutes. So we decided to not use an autoguider for the survey. The
geometric telescope model used for pointing was derived to obtain correction
motions during an exposure, and we also took in account the differential
refraction.

The second problem was to avoid all the periodic or non-periodic errors
of the previous telescope drive. The problems came mainly from the gears.
We decided to replace them by a new mechanism based on a screw without
play for the slow motions. This solution was used both for the declination

(delta) and hour angle (alpha) slow motion control. The telescope mounting is an equatorial fork, the pointing limits are -12 hours to 12 hours for alpha and -30 degrees to 90 degrees for delta. So nearly all the of sky at Pic du Midi is observable.

As modern CCDs are small, we used the prime focus of the 55-cm F/3 mirror. We built a standard three-lens Wynn corrector to compensate the defects of the primary parabolic mirror. Wynn corrector can be optimized to correct different effects: our requests were to get 80 percent of the energy and 5 microns distortion on the edge of the field. This second aspect is sometimes neglected, in fact this is important for such surveys, especially if we want to use a shift-and-add method. If the distortion is important (a few pixels) the reduction software has to first compute this distortion which is time consuming. Photometry measurement is also concerned by this correction.

For the first light of PICOCAM we used a CCD Kodak KAF 1600 which is a 1600 x 1000 pixels chip. The pixel size is of 9 microns, i.e. 1.1 arcsec on the sky. The field is of 0.15 square degree. The read-out noise is 10 e$^-$, this fact being not a problem because the image noise is rapidly dominated by the background photon noise. The CCD is read in 10 seconds. In the near future we hope to use a CCD KAF 4000 which covers the whole corrected field.

2   Centaurs and KBO Research

The first program of the telescope will be a quick survey of the entire observable sky to find bright distant objects. Each field will be observed three times with an exposure time of 120 seconds. We hope to reach m = 21. With the 200 clear nights at Pic du Midi, we may scan about 10 000 square degrees per year. This represents nearly the whole observable sky if we avoid the Milky Way and the bright star regions. With a standard luminosity function (Jewitt et al. [1]) we can reasonably hope to obtain 500 distant minor bodies. If we keep the KAF 1600 chip, the result will still remains at the level of 70 objects.

As the Kuiper belt inclination distribution is not well known, this quick survey can be useful to explore the region far away from the ecliptic. Another important scientific goal is the search for Pluto-size objects. At 100 AU, the magnitude of a Pluto-like object is about 20. With a typical cometary albedo of 0.04, a Pluto-like object can be discovered at 60 AU.

The second program will be the "deep" PICOCAM survey. This time, we want to reach m = 22.5, so it will be necessary to obtain 18 five-minute exposures for each field. With 200 clear nights, we hope to scan 1000 square degrees. We can find 200 distant minor bodies if we refer to a standard luminosity function and only 15 if we still use the CCD KAF 1600. The second survey is clearly less efficient than the quick one.

3   Comet Survey

As the telescope will be fully automated, some observational programs may be performed as a comet survey. The basic idea is to observe every comet with a magnitude less or equal to 20 every week. The first program is a standard photometry work using B, V and R filters. At the present time, comet magnitude measurements come mainly from amateur astronomers and are much too noisy (residuasl larger than one magnitude). There is also a gap between visual observers and CCD measurements. So, the main goal of this survey is to obtain a homogeneous set of photometrical data. Every day Landolt calibration stars (Landolt [2]) will be observed. It will be also possible to calculate the production rate by comparing with OH radio observations. Secondly we can also observe any comet outburst. At this occasion, we may change our real time observation schedule to follow the comet every day.

The astrometry is the second purpose of the survey. Since photometry measurements are made with many different instruments, the standard residuals are from 0.5 to 1 arcsec. As we will have a large field of view we may use the Hipparcos TYCHO catalogue with one million stars. It is possible to have 25 calibration stars within our field. As the pixel size is of 1.1 arcsec, it is reasonable to have a 100 mas internal error. With our homogeneous set of data we hope to solve the old problem of the difference between the gravity centre and light centre.

Another old problem is the non gravitational forces: the three terms used at the present time are a raw approximation of the physical problem. It will be possible with homogeneous and better measurements to improve the standard model. If we combine the production rate given by photometry and the measurement of the non-gravitational forces, it will be possible in some cases to estimate the nucleus mass.

4   Conclusions

Small telescopes still have a future, but this future is necessarily combined with a high automation level. In these conditions many boring programs which need a lot of time can be successfully performed. Many of these programs will never be possible on the new large telescopes. First of all is a survey of the minor bodies of the outer solar system. The time allocated on large telescopes oblige observers to search only along the ecliptic. Even if our survey is limited to $m = 21$, we hope to find objects far away from the ecliptic to calculate a better thickness of the Kuiper belt. The second PICOCAM purpose is a survey of all comets with a magnitude less or equal to 20. These photometric and astrometric measurements over several years will represent a unique set of homogeneous data. It will be possible to revisit some problems such as magnitude measurements or non gravitational forces model. Depending on Pic du Midi refurbishing problems, PICOCAM will start working in summer 1999.

References

1. Jewitt, D., Luu, J., Chadwick, T. (1998): Kuiper belt objects: The Mauna Kea 8K survey (A.J.), **115**, 2125–2135
2. Landolt, A.U. (1992) A.J. **104**, 340

The Power of Pencil-Beam Searches for Trans-Neptunian Science

Brett Gladman

Observatoire de la Côte d'Azur, Departement Cassini, B.P. 4229, F-06304 Nice Cedex 4, France

Abstract. In order to reach the faintest possible limiting magnitudes for moving objects, observations that shift and combine a large number of brief exposures allow reaching a limit comparable to a single long exposure. Additionally, it is possible to recombine the images at rates corresponding to different candidate orbits, and thus search all possible portions of phase space for objects at different distances, orbital inclinations, etc. If recent steep estimates of the luminosity function of the trans-Neptunian region are correct, this method provides twice as many objects per unit telescope time, and thus is well-suited to quickly exploring the gross morphology of the belt (*e.g.*, the inclination distribution).

1 Introduction

The majority of the known trans-Neptunian objects (TNOs) have been discovered with what shall be referred to below as 'the classical method'. The classical method takes 3 exposures of a region of sky, with a spacing of roughly one hour between the exposures. These 3 exposures are then searched by eye or with an automated computer algorithm for moving objects. The limiting magnitude of this method is essentially the point-source limiting magnitude of the worst of the 3 frames (if conditions vary between the exposures). Rectilinear motion with angular rates less than 10 arcsec per hour and oversampled PSFs serve as criteria for rejection non-TNOs (see Trujillo and Jewitt [4]).

Since typical TNOs move 3–4 arcsec/hour, exposure times are limited to 10–15 minutes in arcsecond level seeing conditions; longer exposures suffer 'trailing losses' since the object moves by more than 1 FWHM during the exposure, thus trailing its signal region out and encompassing more and more noise as the exposure length increases. Thus 2 and 4-meter class telescopes have single-exposure limiting magnitudes ranging from $R = 22$ to 24 in typical conditions, which therefore becomes the limit of the classical search.

2 Going Deep

The trailing problem can be circumvented by use of what might be called the 'pencil-beam' method, or sometimes 'digital tracking'. This method is a variant of what has often been done in the past to search for distant known,

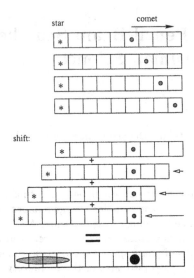

Fig. 1. Schematic representation of the shift and combine technique.

but faint, comets; in that case the telescope can be slewed at the known (non-sidereal) rate of sky motion of the target, resulted in trailed stars but an stationary image of the comet. In TNO searches this of course is impossible since the target's rate of motion is not known. However, suppose that many exposures of the same field are obtained at the sidereal rate, that each of these are short enough that trailing losses are negligible, and that images are taken continuously over a long enough time base that the object moves by many pixels between the first and last exposure. These images can be recombined by shifting each image back with an offset calculated using the time difference between it and a the first image and an assumed rate of motion, exactly cancelling the object's motion (Figure 1). All the stationary objects trail in the direction of recombination, but an object moving at the hypothesized rate will appear stationary in this set of 'shifted' images. The shifted images could simply be summed, but in fact a more useful way to recombine the images is to take a median of the shifted stack. Since the moving object is stationary the median operation will result in an image with that object present; in contrast, stars and galaxies will have moved from their original position. If the target object has moved many FWHM during the total set of observations then the median in principle should remove the stars and galaxies.

In practise, due to the extended 'wings' of typical PSF functions, the object must move by at least 10 FWHM for the median to work well, and 20 or 30 would be desirable. Usually this is limited by the amount of time that a field at opposition is above an airmass of 2, roughly 7–7.5 hours. At 3 arcsec/hour in 1-arcsec seeing, 20 FWHM requires all 7 dark hours available on the opposition field; although there is no reason that multiple nights of data could not be combined in this way. In excellent seeing conditions a long arc would give optimal results. To take the VLT as a specific example, 7 hours on an opposition field in 0.6 arcsec seeing, searching for targets moving 3 arcsec/hour would yield motion of 35 FWHM. In such conditions the recombined limited magnitude for the night's exposure set should be $R \sim 27.5$.

It is tempting to process the pre-shifted images by first subtracting a median image of all the aligned (but unshifted) images. This would in principle yield a set of frames with all stars, galaxies, and other stationary features eliminated, with only moving objects and the inevitable cosmic rays. In the author's experience this method works poorly on ground-based data due to the inevitable variations in seeing during the night; after the subtraction of the median most of the images have the stars replaced with rings where the shoulders of the PSF have not correctly subtracted away. Such images are very difficult to then search. The stability of space-based observing platforms allows this prior subtraction of a 'master' median to work quite well.

Figure 2 shows the final result of such an analysis performed on a set of Palomar 5-meter images. Gladman et $al.$ [1] (G98 hereafter) discusses the results obtained to date for this program at Palomar and CFHT.

3   Future Prospects

Observations have shown that the cumulative number of TNOs brighter than a given apparent R-magnitude limit (that is, the luminosity function of the trans-neptunian belt) appears to have a roughly power-law behavior, given by $\log \Sigma(m_R < R) = \alpha(R - R_0)$, where R_0 is the magnitude where $\Sigma(m_R < R)$ passes 1 TNO per square degree. Luu and Jewitt [3] (LJ98) estimate $\alpha = 0.54 \pm 0.04$ and $R_0 = 23.2 \pm 0.1$, while G98 give $\alpha \simeq 0.76 \pm 0.10$ and $R_0 = 23.4 \pm 0.2$, where the differences are largely due to the different analysis techniques used. This author maintains that a least-squares fit performed on the binned cumulative surface density estimates is formally incorrect, and that the maximum likelihood analysis performed by G98 gives more reliable results. In particular, the error estimate on the slope of the luminosity function produced by a least squares analysis is more or less meaningless, and the value of $\alpha = 0.54 \pm 0.04$ incorrectly implies that the slope is rather well established. Figure 3 shows estimates of the luminosity function.

The number of objects that will be discovered by the classical method (N_c) versus the deep method (N_d) is shown in G98 to be

$$\frac{N_d}{N_c} = \frac{1}{6} 10^{1.6\alpha} . \tag{1}$$

Thus, for either $\alpha = 0.54$ or 0.76 the deep method discovers more objects per unit telescope time than the classical method. In the case of $\alpha = 0.76$ this ratio is almost a factor of 3.

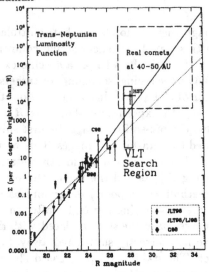

Fig. 2. Estimates of the luminosity function. The surveys can be identified by referring to G98; in particular solid symbols represent surveys with published and detailed detection efficiency functions, used in G98's maximum likelihood analysis. Surveys represented by squares are those summarized in LJ98. Two open triangles represent unpublished surveys by Chiang *et al.* (C98) and Bernstein *et al.* (B98). (The former survey has now been published as Chiang, E., and Brown, M. (1999). Astron. Journal, Astron. J., 118: 1411-1422. These authors report a luminosity function slope in agreement with LJ98 - Eds.). The solid and dashed lines are the best-fit luminosity functions proposed by G98 and LJ98 respectively.

Of course, most objects discovered by the deep method will be very faint, and recovery will be difficult. Thus, it will be very difficult (impossible?) to find track them over the several month period necessary to establish their orbits. So what use is the pencil-beam method? The answer is that the better statistics produced by this method (if the luminosity function is steep) allow one to quickly establish gross morphologies of the belt.

As a specific example of this, consider the inclination distribution. Jewitt, Luu, and Chen [2] claim a broad distribution (Gaussian with FWHM of 30 degrees) is consistent with the data, based on debiasing their ecliptic surveys. Since the surface density out of the ecliptic (at 15 degrees, say) is down by a factor of two, it will require a large amount of 'classical' survey time to find enough objects at each latitude to establish a reliable measure of the relative surface density. Since 1997 the author (with several collaborators) has been attempting to perform a latitude survey using wide-field mosaics on the CFHT (which has recently been plagued with instrumental and hardware problems). The idea is simple, a good night on a 4-meter class telescope should be able to reach $R=26.4$, and at this level the G98 luminosity function predicts $\simeq 200$ objects per square degree. A 30×30 arcmin field mosaic camera should thus detect about 50 objects per field (where each field has been imaged for an entire night). If 4 nights are used to image fields at ecliptic latitudes of 0, 10, 20, and 30 degrees, the fall-off of the number of objects with ecliptic latitude should be easily visible; even that character (Gaussian or not) of the distribution should be reliably determined. This experiment would thus determine the latitude distribution of the belt in a single observing run (at least, at a particular heliocentric longitude). To go from the latitude to the inclination distribution will be somewhat (but not severely) model-dependent, since the one-night TNO observations will not determine the orbital inclinations of the discovered objects. Of course, objects found at latitude x will have minimum orbital inclinations of x degrees; models for this type of problem have already been extensively developed to deal with the IRAF observations of zodiacal dust.

References

1. Gladman, B., Kavelaars, J., Nicholson, P., Loredo, T., Burns, J.A. (1998) Pencil-Beam surveys for faint trans-Neptunian comets. AJ, 116, 2042–2054.
2. Jewitt, D., Luu, J., Chen, J. (1996) The Mauna Kea-Cerro-Tololo (MKCT) Kuiper Belt and Centaur Survey. AJ 112, 1225–1238.
3. Luu, J., Jewitt, (1998) Deep Imaging of the Kuiper Belt with the Keck 10 meter Telescope. ApJ Lett. 502, L91–L94.
4. Trujillo, C., Jewitt, D. (1998) A Semiautomated Sky Survey for Slow-Moving Objects Suitable for a Pluto Express Mission Encounter. AJ, 115, 1680–1687.

Astrometry of Outer Solar System Bodies
Experience with a Small Telescope and Future Plans

Jana Tichá, Miloš Tichý, and Zdeněk Moravec

Klet' Observatory, Zátkovo nábřeží 4, CZ-370 01 České Budějovice, Czech Republic

Abstract. Astrometry of faint and slowly moving objects in the outer solar system is considered to be the target for large telescopes, but in special cases observations using small telescopes were obtained. We have measured precise positions of several Centaur-type asteroids (including recovery of 1997 CU_{26}) and also of the Transneptunian Object 1996 TL_{66} using a 0.57-m reflector equipped with a CCD at the Klet' Observatory. This work will be extended to fainter objects with new 1.02-m reflector at Klet', whose start of operation is planned at the turn of 1998/1999.

1 Follow-Up Astrometry at Klet'

Astrometric measurement of several outer solar system minor bodies were obtained to secure sufficient astrometric data for the determination of reasonable orbits. These observations cover not only the bright Centaurs (2060) Chiron and (5145) Pholus, but also Centaurs (8405)=1995 GO, 1997 CU_{26} and the Transneptunian Object 1996 TL_{66}. We use a 0.57-m f/5.2 reflector equipped with SBIG ST-8 CCD camera. The limiting magnitude is 19.5^m for a 120-second exposure under standard conditions; under very good sky conditions and no moonlight, it is 20.4^m for a 180 second exposure [1].

2 Recovery of Centaur 1997 CU_{26}

This object was discovered by Spacewatch on 1997 February 15. Follow-up astrometric observations were made till 1997 May 28, including 50 observations obtained at Klet' during 13 nights from 1997 February 17 to May 17.

Consequently we selected 1997 CU_{26} as the target for recovery in September 1997, when it appeared on the morning sky at western elongation 50 degrees. The recovery of 1997 CU_{26} was made at Klet' on 1997 September 28.11 UT and confirmed next night on September 29.11 UT [2]. Its brightness was $V = 17.8^m$ and it was found 27 arcseconds from the predicted position. An identification with 1997 CU_{26} was made using a variation of one-opposition orbital elements [3] of this object.

3   Astrometry of 1996 TL$_{66}$

This Transneptunian Object with the highest known eccentricity was discovered on 1996 October 9 by D. Jewitt and colleagues at Mauna Kea, Hawaii, but news of its discovery and preliminary orbit were published as long as the end of 1997 January. Considering its magnitude we selected it as a target for follow-up astrometry. We obtained two positions on 1997 February 1.75 UT and one on February 2.74 UT using 180 second exposures. The total daily motion was 42 arcseconds and its brightness was about 20.7^m V. These observations helped to refine orbit solution of 1996 TL$_{66}$ [7] as well as to enable us testing a technique for astrometric observations of slowly moving and very faint objects with our 0.57-m telescope.

4   The New Project KLENOT

The KLENOT project is the project of the Klet' observatory Near earth and Other unusual objects observations Team (and Telescope) with respect to the fainter objects up to limiting magnitude $V = 22^m$. There are about 160 clear nights per year at Klet' and all observing time will be dedicated to our team. The KLENOT Project goal in outer solar system is to contribute to accurate orbit determination of distant objects. We plan to perform follow-up astrometry and recoveries in further oppositions of Centaurs and brighter Transneptunians up to limiting magnitude $V = 22^m$. All CCD images will be also searched for possible new objects.

The KLENOT telescope:

- 1.02-m primary mirror, primary focus corrector, field of view 0.5 × 0.5 deg.
- CCD camera Photometrics Series 300, chip SITe003B 1024 × 1024 pixels, pixel size 24μm, pixel scale is 1.8 arcseconds per pixel, liquid nitrogen cooling
- this telescope is built with using existing infrastructure of our observatory
- the original mounting was upgraded and the optoelectronical control system was added
- computer equipment, software development in progress
- start of operation is planned in the turn of 1998/1999

5   Acknowledgements

This work has been supported by the Grant Agency of the Czech Republic, Reg. No. 205/96/0042 and Reg. No. 205/98/0266. J. T. and M. T. also thanks the Organizing Committee of ESO MBOSS-98 Workshop for financial support.

References

1. Tichá, J., Tichý, M., Moravec, Z. (1998) Minor planets at Klet' - from discovery to numbering. Planet. Space Sci. **46**, No. 8, 887-891
2. Marsden, B. G. (1997) Minor Planet Electronic Circular 1997-S14
3. Williams, G. V. (1997) Minor Planet Circular No.30094
4. Marsden, B. G. (1997) Minor Planet Electronic Circular 1997-C12

TNO Follow-Up Observations at the Saji Observatory

Atsushi Miyamoto, Hiroki Kosai, Takaaki Oribe

Saji Observatory, 1071-1 Takayama, Saji-son, Yazu-gun, Tottori 689-1312, Japan.
e-mail: sajinet@infosakyu.ne.jp

Abstract. The distant minor-planet-like objects, TNOs (Trans-Neptunian Objects), are very faint (magnitudes mostly below 20). The Saji Observatory started follow-up observations of TNOs in June, 1997, using a 1.03-m reflector, as one of the first programs at that observatory. We wish to contribute to improvement of the TNO orbital elements by this program. We also hope to demonstrate that such activities by the public observatories have great possibilities to contribute to the modern astronomy.

1 Saji Observatory and Its Instruments

Our observatory was established by Saji village, Tottori prefecture, Japan, in July, 1994, as one of about 250 public observatories in our country. Observations are performed every night, except when the weather is bad. More details are available at our web site at http://www.infosakyu.ne.jp/sajinet.

We use a 1.03-m reflector (F/10.2 Cassegrain with an F/4.2 corrective lens with a CCD (STAR-I; 576 x 384 pix;; 23 μ pixel size; 1.07 arcsec/pixel; field size: 10.4 x 6.9 arcmin^2).

There is also a 15-cm F/12 refractor.

2 TNO Observations

A number of TNO's were observed with exposure times from 4 - 120 min and with no filter: 1996 TO$_{66}$ (cubewano, 20.6 mag), 1996 TP$_{66}$ (plutino, 20.8 mag), 1997 CS$_{29}$ (cubewano, 21.4 mag) and 1997 CT$_{29}$ (cubewano, 21.5 mag).

Two to four frames are added and the astrometric measurements are performed with the Astrometrica software (Version 3.0).

We report to MPC of IAU via Mr. Nakano, liaison in Japan.

Although these observations are near the limits of our ability, we try hard to perform follow-up observations of as many TNOs as possible.

Detection of the Small EKB Objects by Occultation with Corot

Françoise Roques

Observatoire de Paris, F-92195 Meudon Principal Cedex, France

Abstract. The size distribution and the spatial distribution, critical clues for the knowledge of the dynamical evolution of the Kuiper Belt, can be reached by the detection of small objects of this population. Stellar occultations are a powerful tool to study dark matter in the Solar System, like asteroids or planetary rings. The possibility to explore the Kuiper Belt by occultation is presented here: Numerous small Kuiper Belt Objects can be detected by high speed photometric observation of well chosen stars. The conditions of these observations are presented, from ground-based telescopes and with the space mission of high precision photometry, COROT.

1 Introduction

A key parameter for the knowledge of the Kuiper belt is the size distribution and in particular the number of comet-sized bodies. These objects are not accessible by the direct observation of the sun-reflected light. In the Solar System, stellar occultations observed with high speed photometry, have led to the discovery of the narrow rings of Uranus and Neptune. The idea to explore the Kuiper belt by occultation, which has been proposed for several years [1,2,5,8], begins to approach reality, thanks to progress made in detector technology. The emerging possibility of fast photometry with CCD camera will certainly make the stellar occultations a fruitful method of exploration of the Kuiper Belt, in particular of the small objects.

2 Stellar Occultations to Explore the Solar System

Stellar occultations have provided a way to reconstruct the shape of asteroids, to explore planetary atmospheres and to discover and carefully map the very narrow rings of Uranus (1 km). The incomplete rings of Neptune (arcs) are much more difficult to be detected than the Uranus rings because the occultation by the arcs is scare and the reality of the detection cannot be confirmed by the symmetric event with respect to the planet. However, the discovery and study of these arcs thanks to stellar occultations shows the feasability of the research by occultation of small objects in the outer solar system and the interest of simultaneous observation with two telescopes to confirm the reality of one detection. In the occultation lightcurves by Uranus and Neptune, isolated events were often observed. One of them, simultaneously observed on the two curves of an occultation by Uranus observed with nearby telescopes, has been analysed as a possible satellite of Uranus of 2 km radius orbiting at 158000 km from the center of the planet [8].

3   The Kuiper Belt Exploration by Occultation

Various parameters affect the detection of Kuiper Belt Objects (KBO) by occultation.

The direction of observation is very important, because it determines the velocity v of the KBO with respect to the star: $v = v_E(cos(\omega) - R^{-1/2})$, where v_E is the Earth velocity, ω is the angle from opposition to the observation, and R is the radius of the KBO orbit in UA. The probability of occultation is proportional to v, but the occultation duration is inversely proportional to this velocity. Toward the opposition, the velocity and the occultation rate are maximum but the occultation duration is minimum. In the direction defined by $cos(\omega) = R^{-1/2}$, the velocity of the KBO in reduced to the velocity perpendicular to the ecliptic plane. In the following of this text, this direction is called "quadrature". This effect allows to distinguish a population of KBO from nearer or farther population (see Table 1).

Table 1.

R(AU)	F(km)	r_{lim}(m)	dt(op) (sec)	ω(quad)	dt(quad) (sec)
4.	0.4	1.5	0.003	$60°$	0.03
40.	1.2	150.	0.02	$82°$	1.
400.	3.8	1500.	0.2	$87°$	31.
4000.	12.	15000.	2.	$89°$	1000.

The star size is a critical parameter because the occultation profile is smoothed on the stellar disc. Typical star radii are between 3.10^{-3} to 1 milliarcsecond. Projected at 40 AU, this corresponds to 0.1 to 30 km radius. Occultation of large stars by comet-sized KBO does not generate a detectable decrase of the stellar flux. Then, the study of comet-sized objects needs stellar candidates carefully chosen. The best candidate stars have a small angular diameter, but are bright enough to preserve a good lightcurve S/N. Observations with a CCD camera allows to record several stars together and then to multiply the number of occultations. The best fields, which contain the maximum of stars near the ecliptic, are at the intersection of the ecliptic and the galactic plane, i.e. in the Sagittarius or the Taurus constellations.

The S/N of the lightcurve limits the detection of the dip. The scintillation is the main limitation on the S/N of the lightcurve for stars brighter than $m_V \approx 12$. At frequency larger than 1 Hz, a S/N of 100 is quite easily obtained, and a S/N of 1000 could be marginally reached with large telescopes.

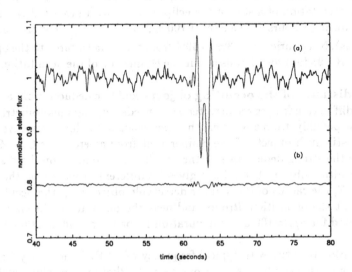

Fig. 1. Synthetic occultation profiles with f=10Hz and v=2km/sec: (a) r(KBO)=0.3 km, r(star)=0.5 km, S/N=100. (b) r(KBO)=0.05 km, r(star)=0.2 km, S/N=1000, shift=-0.2

The diffraction must be taken into account for objects of the order of F, the Fresnel scale, which is the typical scale of the diffraction: $F = \sqrt{(\lambda.R)} \approx$ 1.2 km for KBO observed in visible wavelength λ. The diffraction reduces the depth of the event but extend the size of the diffraction shadow, which is much larger than the geometrical shadow. For more details, see Roques et al. [6] and Roques [7]. KBO smaller than the projected size of the star are detected if the decrease of stellar flux created by the occultation, $\frac{dF}{F} = 2(\frac{r}{r_*})^2$ is detected in the lightcurve. $\frac{dF}{F}$ is proportional to the ratio of the KBO surface to the star surface, the factor 2 is due to a diffraction effect. This perturbation is detected if $\frac{dF}{F}$ is larger than k times the rms fluctuations of the lightcurve. A S/N of 100 allows to detect KBO smaller than the projected stellar diameter (Figure 1-a). For star much smaller that the Fresnel scale, the depth of the profile is reduced with respect to the above formula (see Roques and Moncuquet [7] for more details).

The occultation rate is estimated from a size distribution extrapolated from the known KBO, that is 10^{10} KBO larger than 1 km radius and a differential size distribution index of -4 [4]. The KBO inclination are supposed to follow an exponential distribution with a scale height of 7^0. Then, the probability of occultation of a star of the ecliptic plane with projected radius r_*, observed during 8 hours with a S/N of 100 is $Nocc = 10^{-3}\frac{v}{r_*^2}$. For $r_* = 0.2$ km (0.007 mas), Nocc varies from 3% to 60% from the "quadrature" to the opposition. See RM98 for more details about the estimation of the occultation rate.

The distance of the occulting objects could be deduced from a single event, if diffraction fringes are detected (this needs very high photometric precision), or possibly, from a wavelength dependence of the lightcurve. Another way to distinguish objects of the Kuiper Belt from asteroids or Oort Cloud comets is the strong dependance of the occultation geometry on the distance of the occulting object. The table 1 gives for different orbit radius, the Fresnel scale F, the radius of the smaller detectable object r_{lim}, the occultation duration at the opposition $dt(\text{opp})$ and near the quadrature, $dt(\text{quad})$. r_{lim} is computed for S/N=100 and the duration is for a star radius of 0.02 milliarcseconds.

Fast photometry with typical frequency of 10 Hz is necessary because occultations last usually less than one second. Photometers allow high frequency but they are limited to follow one star. Observations with CCD camera allow to record thousands of stars but limit the acquisition frequency to few Hz.

The Taiwan American Occultation Survey (TAOS) team will deploy a set of small telescopes dedicated to this research along a 7 km east-west baseline in Taiwan. These robotic 50-cm telescopes equipped with CCD camera will automaticaly monitor 3000 stars. They are expected to detect ten to four thousand events per year (see http://www.taos.asiaa.sinica.edu.tw).

Observation with a large telescope would greatly improve the S/N of the lightcurves, which allows to detect much smaller, and then much more numerous objects: High-speed photometry with an 8-meter telescope could reach a S/N of 1000: This allows to detect 50-meter KBOs (Figure 1-b) The above size distribution leads to an occultation rate of 2 to 50 per night. Moreover, the diffraction fringes allow to estimate the distance of the occulting object.

4   Corot

Corot is a French space mission of the CNES dedicated to asteroseismology and to the research on extrasolar planets. For these aims, Corot will perform high-precision photometric observation of stars with a 25-cm telescope equipped with CCD detectors. There are two fields with slightly different optics: One is dedicated to the asteroseismology program and the other one to the extrasolar planets research. The asteroseismology program needs an acquisition frequency of 1 Hz. The two main programs need continuously recording of stars flux during 6 periods of 5 months separated by one month periods which could be dedicated to complementary programs.

Corot presents good characteristics to detect Kuiper Belt Objects by occultations.

The asteroseismology program records few very bright stars. Then, the number of KBOs detected during this program will be limited by the star radius and the small number of stars: the estimated occultation rate will vary from 1 to 100 occultations per month depending on the angle from opposition. KBOs larger than 200 meters can be detected.

The extrasolar planets program records thousands of stars. However, the minimum integration time is 30 seconds. Then, the rate of occultations by KBO is high, but the minimum radius for a detectable KBO is 1 km. The estimated detection rate is from 0.5 to 50 per month.

A program dedicated to the research of KBO would follow 100 stars with a frequency of 1 Hz and could detect 250 events per day. A program of one month would provide enough detections to allow statistical determination of the spatial distribution of the Kuiper Belt.

5   Conclusions

The stellar occultation technique is able to detect small objects of the Kuiper belt and then, to strongly constrain the size distribution. This method does not discover individual objects because it is not possible to follow-up the orbit of the detected objects. However, the belt can be probed as a whole by the cumulative detection of a large number of events. A sufficient number of detections allows also to reconstruct the spatial distribution as the inclination distribution or the existence of possible nest of stability (Lagrangian points).

References

1. Bailey M.E. (1976) Nature **259**, 290–291
2. Brown M.J.I. and Webster R.L. (1997) MNRAS **289**, 783–786
3. Cook, K., Alcock C., Axelrod T. and J. Lissauer (1995) BAAS **27**, 70
4. Dyson F.J. (1992) QJRAS **33**, 45–57
5. Luu, J., Jewitt, D., (1998) Deep Imging of the Kuiper belt with the Keck 10-meter telescope Astrophys. J. L., Ap. J. Lett., 502, L91-94.
6. Roques, F., Moncuquet, M. and B. Sicardy (1987) Astron. J. **93**, 1549–1558
7. Roques, F. and M. Moncuquet (1999), in preparation
8. Sicardy, B., Roques F., Brahic, A., Bouchet P., Maillard J.P. et C. Perrier (1986) Nature **320**, 729–731

Possible Mechanism of Cometary Outbursts

Subhon Ibadov

Institute of Astrophysics, Dushanbe 734042, Tajikistan.

The possibility of transformation of the kinetic energy of high-energy (more than 1 MeV) protons ejected during solar flares into the electrical energy of macroscopic electric double layer in the subsurface region of a cometary nucleus is considered. It is found that at certain conditions, concerning dielectric properties of the nucleus, the energy of the electric field generated during strong solar flares is restricted by discharge potential of the nucleus material. This energy is comparable to the energy of large cometary outbursts. Simulation of the electric discharge mechanism of cometary outbursts in the corresponding technical high-voltage generating device seems a relevant problem.

The author is grateful to the Organizing Committee of the ESO Workshop "Minor Bodies in the Outer Solar System" for invitation to the Workshop and financial support.

References

1. Dobrovolsky O.V., 1966, Comets, Nauka, Moscow.
2. Dorman L.I., L.I. Miroshnichenko, 1968, Solar Cosmic Rays, Nauka, Moscow.
3. Haffner J.,1971, Nuclear Radiation and Protection in Space, Atomizdat, Moscow.
4. Hughes D.W., 1991, Possible mechanisms for cometary outbursts. In: Comets in the Post-Halley Era, Dordrecht, Kluwer, v.2, p. 825-851.
5. Ibadov S., 1996, Physical Processes in Comets and Related Objects, Cosmosinform Publ. Comp., Moscow.
6. West R.M., O. Hainaut, A. Smette, 1991, Post-perihelion observations of P/Halley. III. An outburst at r=14.3 AU, Astron. Astrophys., v. 246, p. L77-L80.

The Uppsala-DLR Trojan Survey
of the Preceding Lagrangian Cloud

Claes-Ingvar Lagerkvist[1], Stefano Mottola[2], Uri Carsenty[2], Gerhard Hahn[2],
Andreas Doppler[3], and Arno Gnädig[3]

[1] Astronomical Observatory, Box 515, S-751 20 Uppsala, Sweden
[2] DLR, Institute of Planetary Exploration, Rudower Chaussee 5, D-12489 Berlin,
   Germany
[3] Archenhold-Sternwarte, Alt-Treptow 1, D-12435 Berlin, Germany

Abstract. The Trojan population is considered to be very numerous but a reliable
estimate of the total number of objects librating around the Jupiter Lagrangian
points is still missing. Furthermore, the true size distribution of the smaller Trojans
is largely unexplored. The ESO Schmidt telescope was used during the apparitions
in 1996, 1997 and 1998 to search for Trojan asteroids by covering a large region of the
sky centered on the preceding Lagrangian point, L4. The Schmidt films were visually
inspected shortly after the observations to identify Trojan candidates. Additional
positions of these candidates were secured with follow-up observations with the
Bochum telescope. During the September 1996 survey, which is nearly complete
down to a limiting V magnitude of about 20, we detected almost 400 suspected
Trojans, the large majority of which are new discoveries. An estimate of the total
number of Trojans based on our still preliminary result from the observation of
1996, indicates a population smaller by a factor three with respect to previous
studies (Shoemaker et al., 1989).

1 Introduction

The first Jupiter Trojan asteroid was discovered in 1906 by Max Wolf in
Heidelberg, but Lagrange had already postulated the existence of such objects
almost 150 years earlier. Initially, the discovery rate of Trojans was quite slow
and it was not until the 70's that the number of Trojans started to increase
significantly, mainly through the efforts by van Houten et al. (1970). However,
the present number of Trojans for which reasonably reliable orbits exist (256
und 175 objects in the L4 und L5 clouds, respectively) is just a small fraction
of the estimated total population. This fact has driven us to initiate a new
survey designed to discover most of the L4 Trojans down to a size limit of
10km, secure their orbits and determine their size distribution.

For this purpose we decided to use the ESO 1m Schmidt in combination
with the 60cm Bochum telescope. The Schmidt telescope was used to cover
a large area in the sky, and the Trojan candidates were discovered by visual
inspection of the films shortly after each night's observations. The discovered
objects were observed on the subsequent night with the Bochum telescope,
equipped with a CCD camera. This allowed us to confirm the discovery,

secure new positions for the future orbit determination, and measure their brightness.

2   Observations

During the 1996 and 1997 apparitions we repeatedly covered an area of about 600 square degrees of the sky surrounding the preceding Lagrangian point, L4. Each film, Kodak 4415, was exposed for 60 minutes with the telescope tracked at sidereal rate, yielding a limiting V magnitude for moving objects of about 20. With heliocentric and geocentric distances close to those of Jupiter, our limiting magnitude translates into an absolute brightness of about 13.5, which roughly corresponds to a size of 10km for the smallest objects we detected. The sky coverage of one film is close to 25 square degrees.

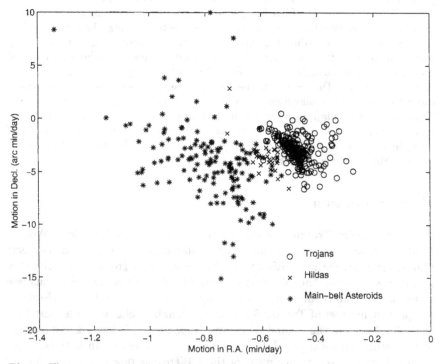

Fig. 1. The apparent motion in Declination and Right ascension of main-belt asteroids, Hildas and Trojans in the observed field.

The idea was to expose each field twice, with a few nights' interval, but bad weather prevented this in some cases. During both apparitions the procedure was repeated about one month after opposition. During the September and

October 1996 campaigns a total of 95 films were taken, while during October and November 1997 the weather conditions were worse and only 72 films were taken. During the 1998 apparition a field of about 1000 square degrees, mainly south of L4 because of the northern declination, was covered with observations by the ESO Schmidt telescope.

Candidate Trojans were identified from their short trails compared to main-belt asteroids. With this procedure, it can happen that asteroids of the Hilda dynamical group are mistakenly flagged as Trojans. We estimate that this spurious component is below the 20% level for each individual observing run. However, linkage of the asteroids positions obtained during different observing runs will enable us to properly identify Hilda asteroids, thus reducing the spurious identifications to a negligible amount.

The 60cm Bochum telescope observations were performed with the DLR MII CCD camera system with a field of view of 9x9 arcminutes. Each field was imaged twice, with an exposure time of 160 seconds, in order to identify the moving objects. Most of the fields were observed again during several nights, to extend the time base for the orbit determination.

3   Astrometry

The CCD astrometry was done using the synthetic aperture software package ASTPHOT, developed by Stefano Mottola. All of the positions on the photographic films from September 1996 were measured directly with the Optronics measuring machine at ESO in Garching. The remaining films from 1996 were digitized with a document scanner and astrometrically reduced with ASTPHOT.

The HST Guide Star Catalog (Lasker et al., 1990) was used as astrometric reference. Normally, 7 - 10 GSC stars were identified interactively, and elliptical apertures were selected for the trailed asteroid images. In this way we determine the centroid of the asteroid, which is assumed to correspond to the position of the object at the midtime of the exposure. Three measurements were performed for each asteroid and the derived astrometric positions were taken as the mean of these measurements.

4   Orbit Determination and Linkage

It is of course our aim to identify as many of the Trojan candidates as possible, either with known asteroids, or to try to determine their orbits and possibly find linkages with astrometric positions previously published. All published positions and orbital elements from the database of the Minor Planet Center are used. So far the linking process has been tested on a subset of the observations from 1997. The linkages are computed with an automated identification search engine developed by Andreas Doppler and Arno Gnädig. It contains the data base of observations, orbital elements, known identifications and routines to calculate orbits. These routines take into account the perturbations of all major planets using the JPL DE403 ephemeris (Standish et al., 1995), and is based on an Adams-Bashforth-Moulton multiple step integrator with variable order and stepsize described in Shampine and Gordon (1975). The orbit corrections were computed using the method developed by Sitarski (1971).

The following procedure is applied:

1) assign observations to numbered asteroids, when possible

2) assign observations to multi-opposition asteroids, when possible

3) find linkages with single-opposition asteroids to determine new orbital elements

4) find linkages with asteroids that do not have initial orbital elements and try to determine elements

5) identify Trojan objects

The examined subset for the first linking contained 255 astrometric positions (from 1997) for 91 objects, including some fast moving objects. It was possible to identify 13 objects with numbered or multi-opposition Trojans. Seven new single-opposition Trojans were found, for 3 of which new orbital elements could be determined. For 5 objects identifications with known main belt asteroids were found. The remaining 63 objects could not be linked and remain in the database. They will be routinely checked as soon as new observations from the MPC are available. For future linkings the database will be expanded to include all available positions from the survey as soon as they have been processed.

5   Results

To date, the data reduction of the September 1996 campaign has been completed. In these fields we discovered a total 399 Trojan candidates. By comparing the relative abundances of previously known Trojans and Hildas in the observed fields, we estimate that roughly 80% of the discovered candidates are actually Trojans. This value is very similar to previous estimates performed by Shoemaker et al. (1989). Furthermore, by assuming that the known Trojans well represent the total Trojan distribution (Fig. 2), we estimate that our fields in September 1996 covered about 55% of the total population, down to a limiting V magnitude of 20, corresponding to an absolute magnitude of 13.5. This estimate implies that the total number of objects in the L4 cloud is about 580 down to a diameter of approximately 10km.

Our result is summarized in Fig. 3. The graph is also showing the cumulative distribution of all Trojan discoveries published as of 1997 along with a previous estimate by Shoemaker et al. (1989). It is worthwhile mentioning that the Shoemaker estimate was based on a much smaller sample. From this graph it can be seen that the Shoemaker estimate (which correspond to

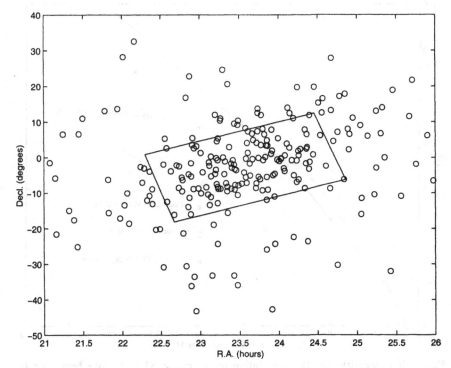

Fig. 2. The position on the sky of known Trojans during the opposition in 1996. The sky coverage is indicated by the parallelogram.

about 1600 objects larger than 10km) is larger by a factor of almost three compared to ours. Even though the correction procedure we applied to determine the total number of Trojans is not very sophisticated, the difference seems to be too large to be reconciled by a more refined analysis. Therefore our conclusion is that the total number of Trojans in the L4 cloud is smaller than previously believed.

A numerical study (in the same way as made for the Hilda group, Dahlgren (1997)) is being performed to understand the dynamical properties of the Trojan population, and interpret them in terms of collisional evolution.

As soon as the positions from the following campaigns will become available, it will be possible to firmly identify the Hilda component in our sample, therefore improving the reliability of our estimate of the size of the Trojan population. Furthermore, for many of the objects it will be possible to compute reliable orbits which will allow us to investigate their dynamical properties, like orbital stability and the presence of dynamical families.

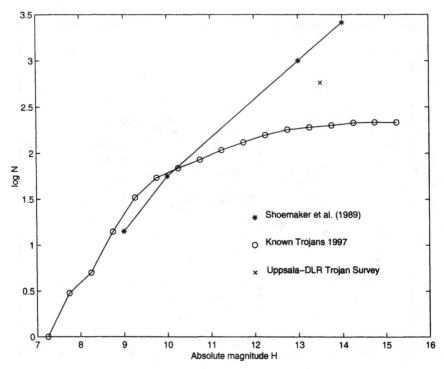

Fig. 3. The logarithm of the cumulative number (log N) of Trojans belonging to the preceding cloud plotted versus absolute magnitude (H). Three sets of data are shown in Fig. 3, from Shoemaker et al. (1989), for all known Trojans at L4 as of 1997 (Minor Planet Center (1998)), and the result from the present survey. For the last we only give the total cumulative number down to our limiting magnitude.

6   Acknowledgements

The observations were carried out under ESO proposals 57F.-0370 and 59F.-0197.

References

1. Dahlgren, M. (1997) A Study of Hilda Asteroids, PhD thesis, Uppsala University.
2. Lasker, B.M., Sturch, C.R., McLean, B.J., Russell, J.L., Jenkner, H., Shara, M.M. (1990) *Astron. J.* **99**, 2019.
3. Minor Planet Center (1998)
   http://cfa-www.harvard.edu/iau/lists/JupiterTrojans.html
4. Shampine, L.F. and Gordon (1975). In Computer solutions of ordinary differential equations. Freeman and Comp., San Francisco.
5. Shoemaker, E.M., Shoemaker, C.S., Wolfe, R.F. (1989) Trojan Asteroids: Populations, Dynamical Structure and Origin of the L4 and L5 Swarms. In Asteroids

II, R.P. Binzel, T. Gehrels, M.S. Matthews, eds., Univ. of Arizona Press, pp. 487–523

6. Sitarski, G., (1971) *Acta Astronomica* **21**, 87.

7. Standish, E.M.; Newhall, X X,; Williams, J.G.; and Folkner, W.M. (1995). "JPL Planetary and Lunar Ephemerides, DE403/LE403". JPL IOM 314.10-127

8. van Houten, C.J., van Houten-Groeneveld, I., Gehrels, T. (1970), Astron. J. **75**, 659-662.

The Distant Satellites of Uranus and the Other Giant Planets

Brian G. Marsden[1], Gareth V. Williams[1], and Kaare Aksnes[2]

[1] Harvard-Smithsonian Center for Astrophysics, Cambridge MA 02138, USA
[2] Institute of Theoretical Astrophysics, N-0315 Blindern, Norway

Abstract. The steady improvement in our knowledge of the orbits of the two new Uranian satellites is described, including the roles of the 1984 prediscovery images and the 1998 recovery observations. The recognition of distant satellites of the other giant planets, notably the lost possible jovian satellite of 1975, is also discussed.

1 Introduction

In early October 1997, B. J. Gladman reported to the Central Bureau for Astronomical Telegrams (CBAT) observations by P. D. Nicholson, J. A. Burns, J. J. Kavelaars and himself of two faint objects comoving with Uranus. The observations, made with the 5-m reflector at Palomar on two consecutive nights a month earlier, were of point sources of red magnitude 21.9 and 20.4, respectively, located some 6 arcmin east and 7 arcmin west-northwest of the planet. Computations at the Central Bureau allowed W. B. Offutt to detect the brighter object at his private observatory in New Mexico on several nights during October, while at the end of that month both objects were briefly recorded both by the discovery group and by D. J. Tholen at the University of Hawaii. Details of the early observations have been published elsewhere (Gladman et al. 1997, 1998).

2 Initial Orbit Computations

Although the concentration of the observations of the fainter object into what were effectively just two points did not allow one to conclude whether this object was orbiting Uranus or not, the better distribution of data made it clear that the brighter object was indeed a satellite—but one could not say with certainty whether it orbited Uranus in a direct or a retrograde sense. Accordingly, the brighter object was given the provisional designation S/1997 U 2, and since it seemed rather likely that the fainter object was also a Uranian satellite, this was given the designation S/1997 U 1.

The initial computation of the orbit of a distant planetary satellite can be facilitated by making trial solutions for various assumed distances from the observer and the constraint that the object is at pericenter or apocenter. One might say that such a process (Väisälä 1939) is essential when the observations are contained in only two points, but it can also be usefully applied in more general situations of indeterminate orbit computation for objects in the outer solar system (Marsden 1999).

Table 1. Apsidal orbits for S/1997 U 2

ρ	e	i	R	ρ	e	i	R	ρ	e	i	R
18.82 P 0.98	169	1+ 7+		19.00 A 0.64	140	4+ 4−		19.14 A 0.14	16	15− 3+	
18.84 P 0.78	168	0 7+		19.01 A 0.72	128	9+ 7−		19.16 A 0.02	15	17− 3+	
18.86 P 0.58	167	1− 7+		19.02 A 0.79	105	14+11−					
18.88 P 0.39	167	2− 7+		19.03 P 0.02	135	19+11−		19.18 P 0.10	14	18− 3+	
18.90 P 0.17	165	3− 6+		19.04 A 0.79	71	13+11−		19.21 P 0.27	13	19− 3+	
18.92 P 0.02	164	4− 6+		19.05 A 0.73	51	6+ 7−		19.24 P 0.43	13	21− 2+	
18.94 A 0.15	162	4− 5+		19.06 A 0.66	39	0 4−		19.27 P 0.58	13	23− 2+	
				19.07 A 0.58	32	4− 1−		19.30 P 0.73	13	24− 1+	
18.96 A 0.32	158	4− 4+		19.08 A 0.52	27	8− 1−		19.33 P 0.86	13	26− 1+	
18.97 A 0.40	156	3− 3+						19.36 P 0.99	13	27− 1+	
18.98 A 0.48	152	2− 2+		19.10 A 0.39	21	11− 2+					
18.99 A 0.56	147	0 0		19.12 A 0.26	18	14− 3+					

Table 1 shows the result of applying the procedure to the observations of S/1997 U 2 on 1997 Sept. 6 and Oct. 31. Inspection of the ephemeris shows that Uranus was at a geocentric distance of about 19.03 AU at the time of the first observation. The need for the orbit about Uranus to be an ellipse constrains the distance ρ of the satellite to the range 18.82–19.36 AU. The column e gives the orbital eccentricity, the P or A preceding it denoting whether the orbit is pericentric or apocentric. The column i denotes the orbital inclination (in degrees and referred to the ecliptic), the increasing distance yielding a general trend from retrograde to direct orbits; there is a circular retrograde solution near $\rho = 18.92$ AU and a circular direct solution near $\rho = 19.16$ AU, while in between the orbits are generally apocentric, except for a brief and ill-defined transition through planet-grazing orbits to a low-eccentricity pericentric case at $\rho = 19.03$ AU. The column R shows the O−C residuals (in arcsec separately in right ascension and declination) of the Oct. 9 observation from each orbit. Inspection of R shows that the most probable solution is therefore the retrograde one near $\rho = 18.99$ AU, although consideration might also be given to a direct orbit near $\rho = 19.06$ AU.

For the purpose of an initial publication (Marsden and Williams 1997a), a restricted least-squares solution similar to the $\rho = 18.99$ AU case was selected; this orbit had $e = 0.52$ and $i = 149$ deg, together with orbital semimajor axis $a = 0.0384$ AU and orbital period $P = 415$ days. Rather arbitarily, a retrograde solution ($i = 138$ deg) with the same values of a and P was adopted for S/1997 U 1, the observations of which extended to Oct. 29; with $e = 0.16$, it seemed already at this time that the orbit of the fainter satellite was significantly more circular than that of the brighter.

A month later, when the observations of S/1997 U 2 extended to Nov. 28 and those of S/1997 U 2 to Nov. 25, slightly modified orbits were published (Marsden and Williams 1997b) in which the values of e were arbitrarily taken to be 0.20 and 0.40, respectively, that for S/1997 U 1 being estimated the maximum possible and that for S/1997 U 2 the minimum.

3   Multiple-Opposition Orbit Computations

Around the end of 1997, Gladman and Burns located possible images of both satellites on photographic plates taken of the Uranus region by D. P. Cruik- shank with the Canada-France-Hawaii telescope on 1984 June 1 and 2. The measurements were only semiaccurate, however, and the identity of the im- ages with the satellites was therefore considered to be inconclusive. Even for the better-observed brighter satellite, it was not clear how many revolutions N the object would have made about Uranus during the 13-year hiatus. Trial orbit fits, in which the perturbations by the sun had to be considered, seemed to suggest that $N = 6$ in the case of this satellite, and a linked orbit was pub- lished (Marsden et al. 1998a) in the hope that it would improve the object's predicted position after the solar conjunction in early 1998. It seemed that $N = 6$ was preferable to a choice of $N = 5$ or $N = 7$ (each of these fits yielding significantly systematic trends in the 1997 residuals), even though the resulting $e = 0.34$ was smaller than the previously suggested minimum of 0.40.

At its recovery in March 1998 S/1997 U 2 was located fully 30 arcsec from this prediction, a circumstance indicating either that the 1984 observations were not of this object or that N was somehow significantly smaller than had been supposed. Further investigation showed that, indeed, there was another solution corresponding to $N = 3.8$, this having been missed earlier because of the larger solar perturbations involved. This solution (Marsden et al. 1998b), fully confirmed by further observations later in 1998, raised the eccentricity again to $e = 0.51$ and refined the inclination to $i = 153$ deg. More significantly, it gave the much higher values of $a = 0.0816$ AU and $P = 1289$ days, this period being the longest known for any planetary satellite. The difference between the pair of 1984 residuals amounted to 2.4 arcsec in right ascension and 4.2 arcsec in declination.

No prediction using the 1984 data was published for S/1997 U 1, and the March 1998 recovery observations were some 3 arcmin from the prediction with maximum assumed $e = 0.20$. The recovery conclusively showed that the eccentricity had less than half this value, as well as that Gladman's 1984 candidate was correct. The linked 1984–1998 solution (Marsden et al. 1998c) had $e = 0.08$, $i = 140$ deg, $a = 0.0479$ AU and $P = 579$ days. The difference between the pair of 1984 residuals was 2.8 arcsec in right ascension and only 0.5 arcsec in declination.

Table 2 gives the latest orbital elements, the symbols n, M, ω and Ω denoting the mean daily motion, mean anomaly at the epoch, argument of the pericenter and longitude of the ascending node along the ecliptic. These results are based on 43 and 119 observations of the respective satellites, the latest in each case made by Offutt in mid-November 1998.

Table 2. Osculating orbits for equinox 2000.0, epoch 1999 Jan. 22.0 TT

	S/1997 U 1	S/1997 U 2
e	0.0815762	0.5113970
a (AU)	0.0478947	0.0813779
n (deg)	0.6213368	0.2805428
M (deg)	172.90892	163.93313
ω (deg)	338.79276	17.60047
Ω (deg)	175.06127	255.57526
i (deg)	139.69174	152.65627

4   Discussion

With the discovery of the new satellites it follows that all four giant planets have the characteristic of possessing distant satellites. It is generally supposed that these satellites have been captured by their primaries, a possibility seemingly rendered more likely by the recognition that the space surrounding these planets is populated by the centaurs (Kowal et al. 1979), objects presumably representing a transition between membership in the Kuiper Belt and in Jupiter's family of short-period comets. Nevertheless, with the possible exception of the four outermost satellites of Jupiter, all twelve of these satellites are tightly bound to their primaries (more tightly than the moon is bound to the earth, in fact), and the captures must have occurred quite early in the lifetime of the solar system, in a collisional environment quite different from what is seen today.

The first ten confirmed distant planetary satellites were found between 1898 and 1974, the earliest being Saturn IX (Phoebe) and the latest Jupiter XIII (Leda). Two of the present authors also participated in the orbit computations for Leda (Kowal et al. 1975a, Aksnes 1978). For those computations the perturbations by the sun, Saturn and the earth were allowed for by subtracting the effect of the perturbations from the observations and using the observations adjusted thereby for a jovicentric two-body least-squares differential correction. In the present case the computation of the uranicentric orbits was done both by carrying out the corresponding two-body procedure and by numerically including the perturbations in the partial derivatives for the differential correction.

The remark was made (Kowal *et al.* 1975a) that there appeared to be an even chance that each of the following oppositions of Jupiter would yield another satellite comparable in brightness to Leda. Although a suspected satellite, now known as S/1975 J 1, was recorded at the very next opposition, there has not subsequently been even a suggestion of a subsequent discovery of a distant jovian satellite. That suspected satellite remains a mystery to this day. The observations, on four nights covering a six-day arc, are incompatible with a heliocentric orbit. Although it was remarked at the time (Kowal *et al.* 1975b) that it was not clear whether the jovicentric orbit was direct or retrograde, direct solutions do in fact give a significantly better fit. A solution that resembles the orbits of Jupiter VI, VII, X and XIII is certainly very plausible, although i seems to be more like $45°$ than in the range $25°$–$30°$ common to those satellites. Nevertheless, despite a careful search, S/1975 J 1 specifically failed to show near the expected position on photographic plates taken two months after discovery. This gives rise to the obvious speculation that the object represented the temporary flaring of a comet physically close enough to Jupiter that the planet would have a pronounced gravitational effect on its motion, possibly with the comet temporarily captured as a satellite (Marsden 1962, Rickman 1979).

References

1. Aksnes, K. (1978): The motion of Jupiter XIII (Leda), 1974–2000. AJ **83**, 1249–1256.
2. Gladman, B. J., Nicholson, P. D., Burns, J. A., Kavelaars, J. J., Offutt, W. B., Tholen, D. J. (1997): Satellites of Uranus. IAU Circ. No. 6764.
3. Gladman, B. J., Nicholson, P. D., Burns, J. A., Kavelaars, J. J., Marsden, B. G., Williams, G. V., Offutt, W. B. (1998): Discovery of two distant irregular moons of Uranus. Nat **392**, 897–899.
4. Kowal, C. T., Aksnes, K., Marsden, B. G., Roemer, E. (1975a): Thirteenth satellite of Jupiter. AJ **80**, 460–464.
5. Kowal, C. T., Roemer, E., Daniel, M. A., Aksnes, K., Marsden, B. G. (1975b): Probable new satellite of Jupiter. IAU Circ. No. 2855.
6. Kowal, C. T., Liller, W., Marsden, B. G. (1979): The discovery and orbit of (2060) Chiron. In *Dynamics of the Solar System* (R. L. Duncombe, Ed.), pp. 245–250.
7. Marsden, B. G. (1962): Violent changes in the orbit of Comet Oterma. Leaflet Astron. Soc. Pacific No. 398.

8. Marsden, B. G. (1999): Kuiper Belt: securing adequate orbital data. In *Asteroids, Comets, Meteors 1996* (A. C. Levasseur-Regourd and M. Fulchignoni, Eds.), in press.

9. Marsden, B. G., Williams, G. V. (1997a): Satellites of Uranus. IAU Circ. No. 6765.

10. Marsden, B. G., Williams, G. V. (1997b): Satellites of Uranus. IAU Circ. No. 6780.

11. Marsden, B. G., Williams, G. V., Aksnes, K. (1998a): S/1997 U 2. IAU Circ. No. 6834.

12. Marsden, B. G., Williams, G. V., Aksnes, K. (1998b): S/1997 U 2. IAU Circ. No. 6869.

13. Marsden, B. G., Williams, G. V., Aksnes, K. (1998c): S/1997 U 1. IAU Circ. No. 6870.

14. Rickman, H. (1979): Recent dynamical history of the six short-period comets discovered in 1975. In *Dynamics of the Solar System* (R. L. Duncombe, Ed.), pp. 293–298.

15. Väisälä, Y. (1939): Eine Einfache Methode der Bahnbestimmung. Astron.-Optika Inst. Univ. Turku Informo No. 1.